AF566679

ISNM 69:
International Series of Numerical Mathematics
Internationale Schriftenreihe zur Numerischen Mathematik
Série internationale d'Analyse numérique
Vol. 69

Birkhäuser Verlag
Basel · Boston · Stuttgart

Numerical Treatment of Eigenvalue Problems Vol. 3

Workshop in Oberwolfach, June 12–18, 1983

Numerische Behandlung von Eigenwertaufgaben Band 3

Tagung in Oberwolfach, 12.–18. Juni 1983

Edited by
Herausgegeben von

J. Albrecht
L. Collatz
W. Velte

1984

Birkhäuser Verlag
Basel · Boston · Stuttgart

Editors/Herausgeber

J. Albrecht
Technische Universität
Clausthal
Institut für Mathematik
D–3392 Clausthal-Zellerfeld (FRG)

L. Collatz
Universität Hamburg
Institut für
Angewandte Mathematik
Bundesstrasse 55
D–2 Hamburg 13 (FRG)

W. Velte
Institut für Angewandte
Mathematik und Statistik
der Universität
Am Hubland
D–8700 Würzburg

CIP-Kurztitelaufnahme der Deutschen Bibliothek

Numerical treatment of eigenvalue problems: workshop ... = Numerische Behandlung von Eigenwertaufgaben. – Basel; Boston; Stuttgart: Birkhäuser
Bis Bd. 2 (1979) u.d.T.: Numerische Behandlung von Eigenwertaufgaben

Vol. 3. Workshop in Oberwolfach, June 12 – 18, 1983. – 1984.
(International series of numerical mathematics; Vol. 69)
ISBN 3-7643-1605-5

NE: PT; GT

Printed in Switzerland by Birkhäuser AG, Graphische Unternehmen, Basel
ISBN 3-7643-1605-5

PREFACE

This volume contains the manuscripts of talks given at a conference on "Numerical Treatment of Eigenvalue Problems" held in June 1983 at the Mathematical Research Institute, Oberwolfach.

The meeting centered in part of methods for the computation of eigenvalue bounds with reports on new developments of various methods (method of Lehmann-Maehly, methods of intermediate problems, estimates using difference methods, inclusion by quotients theorems); numerous examples arising out of physics and engineering sciences (computation of rolling frequences, eigenfrequences of plates and membranes, buckling values, energy levels of atoms and critical values in flow problems) helped to illustrate the significance of these results. Some emphasis was also laid on matrix eigenvalue problems; next to theoretical results (e.g. perturbation theorems), numerical algorithms for the treatment of such problems were discussed, in particular in the form arising from the use of finite elements. There were also reports on problems with nonlinear operators occuring in the treatment of nonlinear oscillations (Duffing differential equation) and in the physics of plasma.

As always in the stimulating atmosphere of the Oberwolfach institute, extensive discussions contributed to intensify the flow of ideas between the participants, among whom a number of guests from foreign European countries and America.

The hearty thanks of the participants go to the director of the Mathematical Research Institute, Herrn Prof. Dr. M. Barner, who, at a time of increasing importance of numerical and applied mathematics, made it possible to held this meeting; to his collaborators in Oberwolfach and Freiburg; and to Birkhäuser Verlag for the excellent presentation of the volume.

Julius Albrecht, Clausthal-Zellerfeld

Lothar Collatz, Hamburg

Waldemar Velte, Würzburg

V O R W O R T

Der Band enthält Manuskripte zu Vorträgen, die auf einer Tagung am Mathematischen Forschungsinstitut Oberwolfach über "Numerische Behandlung von Eigenwertaufgaben" im Juni 1983 gehalten wurden.
Einer der Schwerpunkte der Tagung lag bei den Verfahren zur Berechnung von Eigenwertschranken. Berichtet wurde über Weiterentwicklungen verschiedener Methoden (Lehmann-Maehly-Verfahren, Methoden der "intermediate problems", Abschätzungen mit Hilfe von Differenzenverfahren, Quotienten-Einschliessungssätze); die Bedeutung der hierbei erzielten Resultate wurde an Hand zahlreicher Beispiele aus der Physik und den Ingenieurwissenschaften (Berechnung von Schlingerfrequenzen, von Schwingungsfrequenzen von Platten und Membranen, von Beulwerten, von Energieniveaus von Atomen und von kritischen Werten bei Strömungsproblemen) erläutert.
Ein weiterer Schwerpunkt lag bei den Eigenwertaufgaben mit Matrizen; hier wurden neben theoretischen Resultaten (z.B. Störungssätzen) insbesondere numerische Algorithmen zur Behandlung solcher Matrix-Eigenwertaufgaben, wie sie bei der Verwendung der Methode der finiten Elemente auftreten, besprochen.
Ferner wurde über einige Aufgaben mit nichtlinearen Operatoren, die bei der Behandlung nichtlinearer Schwingungen (Duffingsche Differentialgleichung) und in der Plasmaphysik vorkommen, berichtet.
Wie immer trugen in der anregenden Atmossphäre des Oberwolfacher Instituts auch ausführliche Diskussionen wesentlich dazu bei, den Gedankenaustausch zwischen den Tagungsteilnehmern, unter denen sich zahlreiche Gäste aus dem europäischen Ausland und aus Amerika befanden, zu intensivieren.
Der herzliche Dank der Teilnehmer gilt dem Direktor des Mathematischen Forschungsinstitut, Herrn Professor Dr. M. Barner, der es in einer Zeit wachsender Bedeutung der Numerischen und Angewandten Mathematik ermöglichte, diese Tagung durchzuführen, sowie seinen Mitarbeitern in Oberwolfach und Freiburg, und dem Birkhäuser Verlag für die wie stets sehr gute Ausstattung des Bandes.

Julius Albrecht
Clausthal-Zellerfeld

Lothar Collatz
Hamburg

Waldemar Velte
Würzburg

I N D E X

ISNM, Vol. 69
Numerical Treatment of
Eigenvalue Problems, Vol. 3

THE COMPUTATION OF CONVERGENT LOWER BOUNDS IN QUANTUM MECHANICAL EIGENVALUE PROBLEMS

Christopher Beattie

A computational approach guaranteeing convergent lower bounds to eigenvalues of a wide class of Schrödinger operators is introduced. After motivating the necessary strategy in the simple two-particle case, a variant of Fox's intermediate Hamiltonian construction is presented for complex atoms, together with the necessary convergence criteria. Computational features of Fox's method are discussed in the context of sparse matrix strategies that can utilize the algebraic structure of the resulting matrices to advantage.

§1. Introduction. An important theoretical and practical problem in quantum mechanics is the computation of Schrödinger operator eigenvalues to very high accuracy. Schrödinger operators govern the energy of atoms and atomic ions in accordance with conjectured physical principles describing the interactions among the constitutive particles. Eigenvalues of such operators are related to atomic energy levels, which may be determined quite accurately from experimental spectroscopic data. Comparison with computed eigenvalues for the modeling operator then provides a measure of how accurately the operator models the actual atomic system. The computed eigenvalues themselves are useful in determining whether electronic shell configurations exist and are in fact determined by the operator. Unfortunately it is rarely possible to find exact values for the operator eigenvalues explicitly, so it is necessary to consider computational schemes for obtaining accurate estimates instead.

The problem of assessing the accuracy of an estimate to any given eigenvalue is evidently equivalent to the problem of obtaining rigorous upper and lower bounds to that eigenvalue. Computational procedures that provide rigorous upper bounds are well known and highly refined at this point, e.g. Rayleigh-Ritz (configuration interaction) and Hartree-Fock methods. In contrast, lower bound methods are yet in somewhat of a germinal state. Even so, sufficient knowledge and computational experience has been gained in the last twenty years to indicate a much more complex situation requiring much more subtle analysis and somewhat more computational effort than appears to be required in complimentary upper bound methods.

§2. The Class of Problems to be Treated. We consider the task of estimating the bound state energies of atomic systems having m identical particles interacting with each other and with an infinitely massive nucleus. The model Hamiltonian is taken as

$$H = \sum_{k=1}^{m} \left[\frac{-1}{2\mu} \Delta_k + W(\underset{\sim}{r}_k) \right] + \sum_{i<j}^{m} V(\underset{\sim}{r}_i - \underset{\sim}{r}_j)$$

operating on a domain dense in $L^2(\mathbb{R}^{3m})$. In this notation $\{\Delta_k\}$ are Laplacians acting in copies of $L^2(\mathbb{R}^3)$; $\{\underset{\sim}{r}_k\}$ are position vectors in $\mathbb{R}^3$; W and V are potentials on $\mathbb{R}^3$ representing nuclear and interparticle forces, respectively; and μ is the particle mass. Generally W and V are simple Coulombic interactions, however an interesting alternative was offered by Thirring [17], and explored extensively by Greenlee and Russell ([12], [16]). There, W is taken as an effective screening potential so that V becomes the residual correction needed to regain the exact Coulombic energies.

In order to construct convergent lower estimates to the isolated eigenvalues of H, we require the following three assumptions:

1) The potentials V and W are Kato potentials, i.e. members of $L^2(\mathbb{R}^3) + [L^\infty(\mathbb{R}^3)]_\varepsilon$,
2) $V \geq 0$ a.e. in $\mathbb{R}^3$, and
3) the self-adjoint operator corresponding to $\frac{-1}{2\mu}\Delta + W$ in $L^2(\mathbb{R}^3)$ should have discrete eigenvalues below the lowest point of the essential spectrum that are computable to arbitrarily high precision together with their corresponding eigenfunctions.

Some observations may be made immediately with regard to these assumptions. We first note that the self-adjoint operator associated with

$$H_0 = \sum_{k=1}^{m} \left[\frac{-1}{2\mu}\Delta_k + W(\underset{\sim}{r}_k)\right]$$

has discrete lower spectra that is computationally resolvable in terms of the spectra of $\frac{-1}{2\mu}\Delta + W$ by separation of variables. Furthermore, the symmetric operator

$$\hat{H} = \sum_{i<j}^{m} V(\underset{\sim}{r}_i - \underset{\sim}{r}_j)$$

is essentially self-adjoint on a common core with both H_0 and H. Since $\hat{H} \geq 0$, we see that $H_0 \leq H$ (in the usual sense of ordering for symmetric operators, cf. [18]). Hence we may conclude from standard monotone theorems ([18]) that the lower eigenvalues of H_0 are lower bounds to the corresponding eigenvalues of H and that the lowest point of the essential spectrum of H_0 is a lower bound to the lowest point of the essential spectrum of H. The lower bounds provided by H_0

are generally very poor, so we are led naturally to seek out ways to improve them. This is commonly done by approximating the positive definite operator $\hat{H}$.

In 1950 Aronszajn [1] presented in a broader context a systematic method of approximating $\hat{H}$ from below by constructing a bounded positive semi-definite operator $0 \leq \hat{H}^* \leq \hat{H}$. The monotone theorems appealed to above then indicate that the eigenvalues of $H^* = H_0 + \hat{H}^*$ lie intermediate between those of H_0 and those of H, i.e. they are improved lower bounds to the eigenvalues of H.

The construction begins with a family of vectors $\{p_\nu\}$ chosen in $\mathrm{Dom}(\hat{H})$. An associated family of projections $\{P^\beta\}$ may then be defined on $L^2(\mathbb{R}^3)$ as

$$P^\beta u = \sum_{k,\ell=1}^{\beta} (u, \hat{H}p_k) b_{k\ell} p_\ell$$

where $(\cdot\,,\cdot)$ denotes the inner product on $L^2(\mathbb{R}^{3m})$ and $b_{k\ell}$ are the elements of the matrix inverse to the weighted Gram matrix $[(p_i, \hat{H}p_j)]$. One may check directly that $\hat{H}P^\beta$ is a bounded, positive semi-definite operator on $L^2(\mathbb{R}^{3m})$ and that $0 \leq \hat{H}P^{\beta_1} \leq \hat{H}P^{\beta_2} \leq \hat{H}$ for $\beta_1 \leq \beta_2$ as a consequence of Bessel's inequality in the $\hat{H}$ - weighted inner product space associated with $\mathrm{Dom}(\hat{H})$. The operators defined by $H^\beta = H_0 + \hat{H}P^\beta$ satisfy $H_0 \leq H^{\beta_1} \leq H^{\beta_2} \leq H$ for $\beta_1 \leq \beta_2$ hence the eigenvalues of H^β are nondecreasing in β and always lower bounds to the corresponding eigenvalues of H (thus they are "improvable" lower bounds).

§3. Bazley's Special Choice and Convergence. Aronszajn showed that these improvable lower bounds are computable in principle from information available for H_0, however in practical terms, the required computation is generally infeasible. In 1959, Bazley [2] discovered that for a special choice of $\{p_\nu\}$, the spectrum of H^β may be explicitly computed through the diagonalization of a symmetric matrix of order β. In fact, if $\{u_i^0\}$ denote the eigenvectors of H_0 and $\{u_i^0\} \subset \mathrm{Ran}(\hat{H})$, one may choose $p_k = \hat{H}^{-1}u_k^0$. It is then quite easy to see that $\mathcal{U}_\beta = \underset{\beta}{\mathrm{span}}\{u_k^0\}$ is a reducing space for H^β. Restricted to $\mathcal{U}_\beta$, H^β is equivalent to a matrix operator on $\mathbb{R}^\beta$, whereas restricted to $\mathcal{U}_\beta^\perp$, H^β is equivalent to H_0.

Although Bazley's special choice produces a tractable computational problem, this choice of projecting vectors is poor from the viewpoint of convergence. To clarify this point it is useful to review what is currently known regarding the convergence of Aronszajn's method in the presence of essential spectrum.

One may immediately notice that $\mathrm{Ran}(\hat{H}P^\beta)$ is finite-dimensional hence $\hat{H}P^\beta$ is a compact operator on $L^2(\mathbb{R}^{3m})$, implying in turn that $\sigma_{ess}(H^\beta) = \sigma_{ess}(H_0)$ (Weyl's Theorem). Thus the <u>only</u> eigenvalues of H that are accessible to convergent estimates are those that lie <u>below</u> $\inf \sigma_{ess}(H_0)$ (the lowest point in the essential spectrum of H_0).

With this proviso implicit, we may state convergence criteria due to Brown [6] and Greenlee [11] (a third distinct proof may be found in [4]): If $\hat{H}^{-1}$ extends to a bounded operator then it suffices to require that span $\{p_\nu\}$ is dense in $\mathrm{Dom}(\hat{H})$ endowed with the graph norm $(||u||^2 + ||\hat{H}u||^2)^{1/2}$ in order to guarantee convergence of the eigenvalues of H^β to the corresponding accessible eigenvalues of H.

This indicates the sort of result we seek, however $\hat{H}$ will not generally have a bounded inverse or perhaps not even dense range, hence the following criteria is of more use ([4]): If $\hat{H}$ is essentially self-adjoint on a core for H then it is sufficient to require density of $\{\hat{H}p_\nu\}$ in $\overline{\text{Ran}}\,(\hat{H})$ with respect to the L^2-norm.

Since the eigenfunctions of H_0 are not dense in L^2 whenever $\sigma_{ess}(H_0)$ contains a line segment (which includes our case), Bazley's special choice does not satisfy this convergence criteria. In fact, a stronger statement may be made. If an eigenvector of H does not lie in $\overline{\text{span}}\ \{u_k^0\}$, the special choice estimates to the corresponding eigenvalue <u>will not</u> converge to the proper value.

Although the method of special choice has frequent computational advantages, we do not consider it as a reliable method for problems containing significant essential spectrum. We must relax the constraints imposed by computational feasibility so as to allow a choice of projecting vectors $\{p_\nu\}$ that does satisfy the convergence conditions given.

§4. <u>Truncation Including the Remainder</u>. Bazley and Fox [3] noticed that a computationally feasible lower bound problem can be constructed with arbitrary $\{p_\nu\} \subset \text{Dom}\,(\hat{H})$ if a slightly inferior base operator is used. The new base operator is defined in terms of H_0 as a spectral truncation:

$$H_0^n u = \sum_{k=1}^{n} \mu_k^0(u,\ u_k^0)u_k^0 + \mu_{n+1}^0[u - \sum_{k=1}^{n}(u,\ u_k^0)u_k^0],$$

where $\{\mu_k^0, u_k^0\}$ are eigenvalue-eigenvector pairs for H_0. H_0^n is equivalent to H_0 on $\mathcal{U}_n$ and to $\mu_{n+1}^0 I$ on $\mathcal{U}_n^\perp$. Thus H_0^n is a bounded self-adjoint operator and $H_0^n \leq H_0 \leq H$. The interesting feature of H_0^n is that it is a translation by $\mu_{n+1}^0 I$ of a finite-rank operator with range $\mathcal{U}_n$, hence <u>any</u> subspace containing $\mathcal{U}_n$ is a reducing space for H_0^n. If we consider intermediate problems defined by

$$H^{n\beta} = H_0^n + \hat{H}P^\beta \lessdot H_0 + \hat{H}P^\beta \lessdot H,$$

we may obtain improvable lower bounds to the eigenvalues of H by resolving $H^{n\beta}$ on the $(n + \beta)$-dimensional subspace $\mathcal{U}_n \oplus \operatorname{span}_\beta \{\hat{H}p_\nu\}$ which reduces $H^{n\beta}$ for any choice of $\{p_\nu\}$. On the orthogonal complement of this subspace we have $H^{n\beta} = \mu_{n+1}^0 I$. The projecting family $\{p_\nu\}$ may now be chosen to satisfy any density requirement desired, however due to the spectral information for H_0 that was lost in truncation, convergence of estimates to an eigenvalue of H <u>will</u> <u>not</u> occur if the corresponding eigenvector of H does not lie entirely within $\overline{\operatorname{span}}\ \{u_k^0\}$ (the same necessary condition as seen for special choice). We may recover this spectral information in part by approximating the truncation remainder $\tilde{H}_0^n = H_0 - H_0^n$ in a way analogous to the way $\hat{H}$ was approximated: choose $\{q_k\} \subset \operatorname{Dom}(H_0) \setminus \mathcal{U}_n$ and define

$$Q^\alpha u = \sum_{i,j=1}^{\alpha} (u, \tilde{H}_0^n q_i) c_{ij} q_j$$

with $[c_{ij}]^{-1} = [(q_k, \tilde{H}_0^n q_\ell)]$. Then the intermediate operators

$$H^{n\alpha\beta} = H_0^n + \tilde{H}_0^n Q^\alpha + \hat{H}P^\beta$$

provide improvable lower bounds to the eigenvalues of H. They are reduced by the $(n + \alpha + \beta)$-dimensional subspaces $\mathcal{U}_n \oplus \operatorname{span}_\alpha (\tilde{H}_0^n q_k) \oplus \operatorname{span}_\beta (\hat{H}p_k)$ (on the orthogonal complement, $H^{n\alpha\beta} = \mu_{n+1}^0 I$, as before). Aside from the evaluation of inner products, the computational effort involved consists principally of the diagonalization of a matrix of order $n + \alpha + \beta$. Moreover, the projecting families $\{q_k\}$ and $\{p_k\}$ may be chosen to satisfy density conditions sufficient to guarantee convergence of every estimate to the appropriate accessible eigenvalue of H.

A simple example of an appropriate choice of projecting families will be given for estimating helium bound states. The Hamiltonian we consider is given in atomic units by

$$H = [-\tfrac{1}{2}\Delta_1 - \tfrac{1}{2}\Delta_2 - \frac{2}{|\underset{\sim}{r}_1|} - \frac{2}{|\underset{\sim}{r}_2|}] + \frac{1}{|\underset{\sim}{r}_1 - \underset{\sim}{r}_2|} \text{ on } L^2(\mathbb{R}^6)$$

The bracketed term constitutes the hydrogen-like operator H_0 and the last term becomes $\hat{H}$. Since $\hat{H}$ is relatively compact with respect to H_0, $\sigma_{ess}(H_0) = \sigma_{ess}(H)$. Thus every lower eigenvalue of H is accessible to convergent estimates.

In order to construct convergent intermediate Hamiltonians as described above, we must select $\{p_\nu\}$ and $\{q_\nu\}$. Introduce families $\{\phi_k\}$ and $\{\psi_k\}$ in $L^2(\mathbb{R}^3)$ so that each Ψ_k has exponential decay at ∞, $\{\Psi_k\}$ is dense in $L^2(\mathbb{R}^3)$, and $\{\phi_k\}$ is dense in $W^{2,2}(\mathbb{R}^3)$. For example, we may take $\{\Psi_k\}$ as the complete set of Laguerre polynomials and $\{\phi_k\}$ as the set of isotropic Slater orbitals [12]. Define

$$p_{\nu_{k\ell}} = |\underset{\sim}{r}_1 - \underset{\sim}{r}_2| \, \Psi_k(\underset{\sim}{r}_1) \, \Psi_\ell(\underset{\sim}{r}_2)$$

and

$$q_{\nu_{k\ell}} = \phi_k(\underset{\sim}{r}_1)\phi_\ell(\underset{\sim}{r}_2)$$

The resulting families $\{p_\nu\}$ and $\{q_\nu\}$ fulfill the appropriate density conditions given previously hence the corresponding intermediate Hamiltonians provide convergent estimates. The exponential decay condition is included only to assure that each $p_\nu \in L^2(\mathbb{R}^6)$. Polynomial decay of sufficiently high order may be substituted.

§5. Larger Atomic Systems: The Fox Construction. In most cases of interest involving larger atoms ($m > 2$), the standard Aronszajn method is unable to provide convergent estimates. In such cases, it is typical for some or all of the eigenvalues of H to lie above the lowest point of the essential spectrum of H_0. All such eigenvalues are inaccessible to convergent estimates derived from compact approximations to $\hat{H}$, hence it is necessary to consider variants of the Aronszajn idea that utilize noncompact approximations to $\hat{H}$.

Fox [9] discovered such a construction of intermediate Hamiltonians retaining the critical property of being computationally resolvable. The central idea in Fox's construction involves an approximation of $\hat{H}$ that produces intermediate Hamiltonians having the same "ionic" structure as the original Hamiltonian. We write the m-particle space $L^2(\mathbb{R}^{3m})$ explicitly as an m-fold tensor product $L^2(\mathbb{R}^3) \otimes L^2(\mathbb{R}^3) \otimes \ldots \otimes L^2(\mathbb{R}^3)$. If A_0 denotes the operator $\frac{-1}{2\mu} \Delta + W$ on $L^2(\mathbb{R}^3)$ and A_{ij} denotes multiplication by $V(\underset{\sim}{r}_i - \underset{\sim}{r}_j)$ on $L^2(\mathbb{R}^3) \otimes L^2(\mathbb{R}^3)$ (instead of on $L^2(\mathbb{R}^{3m})$) then the original Hamiltonian may be represented as

$$H = \underbrace{A_0 \otimes I_2 \otimes \cdots \otimes I_m}_{\frac{-1}{2\mu}\Delta_1 + W(\underset{\sim}{r}_1)} + \underbrace{I_1 \otimes A_0 \otimes \cdots \otimes I_m}_{\frac{-1}{2\mu}\Delta_2 + W(\underset{\sim}{r}_2)} + \cdots$$

$$\cdots + \underbrace{I_1 \otimes I_2 \otimes \cdots \otimes A_0}_{\frac{-1}{2\mu}\Delta_m + W(\underset{\sim}{r}_m)} + \sum_{i<j}^{m} \underbrace{U_{ij}^{-1}\left[A_{ij} \otimes \left(\bigotimes_{\nu \neq i,j} I_\nu\right)\right] U_{ij}}_{V(\underset{\sim}{r}_i - \underset{\sim}{r}_j)}$$

The operator U_{ij} is a unitary transformation on $L^2(\mathbb{R}^{3m})$ corresponding to the interchange of variables $\underset{\sim}{r}_i \leftrightarrow \underset{\sim}{r}_1$ and $\underset{\sim}{r}_j \leftrightarrow \underset{\sim}{r}_2$. This is introduced as a notational convenience.

Rather than approximate $\hat{H}$ or $V(\underset{\sim}{r}_i - \underset{\sim}{r}_j)$ directly on the m-particle space $L^2(\mathbb{R}^{3m})$, each $V(\underset{\sim}{r}_i - \underset{\sim}{r}_j)$ is approximated separately on the two-particle space in which it acts. Such an approximation may be argued to be more faithful structurally to the original Hamiltonian since the potential energy of any two-particle cluster should be independent of the states of other particles (neglecting second-order correlation effects). The use of a standard Aronszajn-type approximation on $L^2(\mathbb{R}^{3m})$ introduces spurious coupling energies among the particle states.

The mathematical details of the construction of intermediate Hamiltonians and the corresponding reducing spaces is well discussed in [10]. It will suffice for our purposes to describe only the structural similarities with the original Hamiltonian in order to facilitate an intuitive understanding of how the method works.

Following the notation of the previous section, we truncate the single particle operator A_0 on $L^2(\mathbb{R}^3)$ and approximate the truncation remainder as before so that A_0 is replaced with $A_0^n + \lambda_0^n Q^\alpha$. The multiplication operator A_{ij} is approximated on $L^2(\mathbb{R}^3) \otimes L^2(\mathbb{R}^3)$ by $A_{ij}P^\beta$ where P^β is a projection on $L^2(\mathbb{R}^3) \otimes L^2(\mathbb{R}^3)$ constructed with the $L^2(\mathbb{R}^3) \otimes L^2(\mathbb{R}^3)$ inner product weighted with A_{ij}. The intermediate Hamiltonian then takes the form

$$H^{n\alpha\beta} = (A_0^n + \tilde{A}_0^n Q^\alpha) \otimes I_2 \otimes \cdots \otimes I_m +$$

$$I_1 \otimes (A_0^n + \tilde{A}_0^n Q^\alpha) \otimes \cdots \otimes I_m + \cdots$$

$$\cdots + I_1 \otimes I_2 \otimes \cdots \otimes (A_0^n + \tilde{A}_0^n Q^\alpha) +$$

$$\sum_{i<j}^{m} U_{ij}^{-1} \left[A_{ij} P^\beta \otimes \left(\bigotimes_{v \neq i,j} I_v \right) \right] U_{ij}$$

The operator $H^{n\alpha\beta}$ shares some physically relevant properties with the original Hamiltonian H as a consequence of the evident similarity in structure. For instance, the lowest point of the essential specrum of $H^{n\alpha\beta}$ may be closely tied to the lowest point of the spectrum of a 'smaller' operator coresponding to an m-1 particle system - similar to an HVZ theorem (see [5]). This corresponds to an analogous identification of the ionization energy of an atom with the ground state of the next smaller ion.

The construction of Fox's intermediate Hamiltonians requires the selection of two projecting families $\{p_\nu\} \subset L^2(\mathbb{R}^3) \otimes L^2(\mathbb{R}^3)$ and $\{q_\nu\} \subset L^2(\mathbb{R}^3)$. In order to computationally resolve the $H^{n\alpha\beta}$ it seems necessary to mildly constrain the choice of $\{p_\nu\}$ so that $\operatorname{span}_{\beta\cdot\beta} \{A_{ij}p_\nu\} \subset \Pi^\beta \otimes \Pi^\beta$ for some β-dimensional subspace $\Pi^\beta \subset L^2(\mathbb{R}^3)$, with an explicitly known basis $\{\pi_\nu\}_1^\beta$. The $\{p_\nu\}$ can then be determined by a choice of $\{\pi_\nu\}$ from $p_{\nu_{k\ell}} = A_{ij}^\dagger (\pi_k \otimes \pi_\ell)$ $k,\ell = 1,\cdots,\beta$ (the $\dagger$ denotes a pseudoinverse). Now the families $\{q_\nu\}$ and $\{\pi_\nu\}$ may be chosen in such a way as to guarantee convergence of all estimates to the appropriate eigenvalues of H. The criteria are similar to what was imposed for the standard Aronszajn method:

(1) span $\{\pi_\nu\}$ is dense in $L^2(\mathbb{R}^3)$,

(2) span $\{q_\nu\}$ is dense in $W^{2,2}(\mathbb{R}^3)$,

and there may be an additional proviso on $\{\pi_\nu\}$ to insure that $P_{\nu_{k\ell}} \in L^2(\mathbb{R}^3) \otimes L^2(\mathbb{R}^3)$ (e.g. exponential decrease at ∞). The Laguerre polynomials and Slater orbitals mentioned previously are applicable here as well.

The proof of convergence is given in [5] (the special case of Coulombic potentials is detailed in [4]). The critical aspect of the proof lies in showing that $\sigma_{ess}(H^{n\alpha\beta})$ eventually moves up enough to make the lower eigenvalues of H accessible to convergent estimates. The HVZ-type theorem mentioned previously provides the technical machinery to do this.

§6. Computational Considerations. The practical resolution of the spectral problem for the standard Aronszajn construction of intermediate operators is well-documented in [10]. For the intermediate Hamiltonian construction of Fox however, little account has been taken of the special structure of the very large matrix eigenvalue problems that arise. This special structure is a direct consequence of the special structure of the intermediate Hamiltonians $H^{n\alpha\beta}$.

The reducing spaces for $H^{n\alpha\beta}$ are constructed as follows. If $\{(\lambda_i^0, v_i^0)\}$ denote eigenvalue-eigenvector pairs for A_0, we define $M_{n\alpha\beta} = \text{span}\,[\{v_i\}_1^n, \{A_0^n q_j\}_1^\alpha, \{\pi_k\}_1^\beta]$. All reducing spaces for $H^{n\alpha\beta}$ can be characterized in terms of $M_{n\alpha\beta}$ and have the form $N_1 \otimes N_2 \otimes \cdots \otimes N_m$ with each $N_k = M_{n\alpha\beta}$ or $N_k = M_{n\alpha\beta}^\perp$. There are 2^m such subspaces. Due to the isomorphism of interchange of variables there are only $m + 1$ distinct classes of such reducing spaces and these may be indexed according to the number of factors $M_{n\alpha\beta}$ appearing in the product $N_1 \otimes N_2 \otimes \cdots \otimes N_m$. Following Fox [8], a subspace of type r will refer to a reducing space built up as products of r copies of $M_{n\alpha\beta}$ and $m - r$ copies of $M_{n\alpha\beta}^{\perp}$. Evidently there are $\binom{m}{r}$ subspaces of type r for each $r = 0,\ldots,m$.

Let us consider the resolution of the spectral problem for the restriction of $H^{n\alpha\beta}$ to a representative subspace of type r, denoted $H^{n\alpha\beta}|_r$. Subspaces of type r may be decomposed (possibly following an interchange of variables) as $M \otimes M^{\perp}$ where M is a subspace of $L^2(\mathbb{R}^{3r})$ of dimension $(n + \alpha + \beta)^r$ and $M^{\perp}$ is an infinite-dimensional subspace of $L^2(\mathbb{R}^{3(m-r)})$. With the separation of variables suggested by this decomposition, the spectrum of $H^{n\alpha\beta}|_r$ consists of eigenvalues given by

$$\kappa_i + (m-r)\lambda_{n+1}^0$$

where $\{\kappa_i\}$ are the eigenvalues of a matrix eigenvalue problem of dimension $(n + \alpha + \beta)^r$. The eigenvalues of $H^{n\alpha\beta}|_r$ are of infinite multiplicity if and only if $r \neq m$ thus the spectral information obtained from resolving $H^{n\alpha\beta}|_r$ $r = 0,1,\ldots,m-1$ specifies $\sigma_{ess}(H^{n\alpha\beta})$ completely. Notice that since $\inf \sigma_{ess}(H^{n\alpha\beta})$ is of primary interest only the lowest matrix eigenvalue in the first m matrix eigenvalue problems need be computed $(r = 0,1,\ldots,m-1)$. The eigenvalues of $H^{n\alpha\beta}$ of finite multiplicity are found by diagonalizing a matrix of order $(n + \alpha + \beta)^m$. Only those eigenvalues <u>below</u> the computed $\inf \sigma_{ess}(H^{n\alpha\beta})$ can be considered as providing rigorous lower bounds, hence only the lower matrix eigenvalues need be computed.

Let $\{s_k\}_1^{n+\alpha+\beta}$ distinguish a basis for $M_{n\alpha\beta}$. All the matrices required for the resolution of $H^{n\alpha\beta}$ may be generated from knowledge of <u>three</u> matrices:

$$\underset{\sim}{R} = [\, \langle (A_0^n + \overline{A^n}_0 Q^\alpha) s_i, s_j \rangle], \quad \underset{\sim}{S} = [\langle s_i, s_j \rangle], \quad \text{and}$$

$$\underset{\sim}{T} = [\langle A_{12} P^\beta (s_i \otimes s_j),\ s_k \otimes s_\ell \rangle]$$

The inner products $\langle \cdot , \cdot \rangle$ are taken on spaces appropriate to the contents of the bracket. We adopt the notational convention of $\otimes$ with matrices to denote a Kronecker matrix product. The sequence of matrix eigenvalue problems produced are:

r=0 $\quad K = \lambda_{n+1}^0 \quad$ (trivial)

r=1 $\quad \underset{\sim}{R}\underset{\sim}{x} = \kappa \underset{\sim}{S}\underset{\sim}{x}$

r=2 $\quad (\underset{\sim}{R}\otimes\underset{\sim}{S} + \underset{\sim}{S}\otimes\underset{\sim}{R} + \underset{\sim}{T})\underset{\sim}{x} = \kappa(\underset{\sim}{S}\otimes\underset{\sim}{S})\underset{\sim}{x}$

r=3 $\quad (\underset{\sim}{R}\otimes\underset{\sim}{S}\otimes\underset{\sim}{S} + \underset{\sim}{S}\otimes\underset{\sim}{R}\otimes\underset{\sim}{S} + \underset{\sim}{S}\otimes\underset{\sim}{S}\otimes\underset{\sim}{R} + \underset{\sim}{T}\otimes\underset{\sim}{S} +$

$\quad\quad \underset{\sim}{S}\otimes\underset{\sim}{T} + \underset{\sim}{P}_{13}(\underset{\sim}{T}\otimes\underset{\sim}{S})\underset{\sim}{P}_{13})\underset{\sim}{x} = \kappa(\underset{\sim}{S}\otimes\underset{\sim}{S}\otimes\underset{\sim}{S})\underset{\sim}{x}$

etc.

Here $\underset{\sim}{P}_{13}$ is a permutation matrix corresponding to the unitary operator U_{13} operating on $M_{n\alpha\beta} \otimes M_{n\alpha\beta} \otimes M_{n\alpha\beta}$ so that, for example $\underset{\sim}{P}_{13}(\underset{\sim}{R} \otimes \underset{\sim}{R} \otimes \underset{\sim}{S})\underset{\sim}{P}_{13} = \underset{\sim}{R} \otimes \underset{\sim}{S} \otimes \underset{\sim}{R}$. The eigenvalue problem for

$r = 1$ may be resolved adequately by inspection. The matrices required for $r > 3$ follow the same pattern: the left hand matrix consists of a sum of r matrices built up of r-fold products of copies of $\underset{\sim}{S}$ with a single $\underset{\sim}{R}$ and $(r^2 - r)/2$ matrices built up of $(r - 1)$-fold products of copies of $\underset{\sim}{S}$ with a single $\underset{\sim}{T}$, permuted as described above. The right hand matrix consists simply of an r-fold product of copies of

$\underset{\sim}{S}$.

Generalized matrix eigenvalue problems may be reduced to standard matrix eigenvalue problems if a Cholesky factor can be found for the right-hand matrix. Our case is particularly simple, since if $\underset{\sim}{S} = \underset{\sim}{L}\underset{\sim}{L}^t$ is the Cholesky decomposition of $\underset{\sim}{S}$, then the Cholesky factor of an r-fold product of $\underset{\sim}{S}$ is given by an r-fold product of L, e.g.

$$\underset{\sim}{S} \otimes \underset{\sim}{S} \otimes \underset{\sim}{S} = (\underset{\sim}{L} \otimes \underset{\sim}{L} \otimes \underset{\sim}{L})(\underset{\sim}{L}^t \otimes \underset{\sim}{L}^t \otimes \underset{\sim}{L}^t).$$

In considering numerical algorithms for solving these matrix eigenvalue problems one immediately notices that the matricial dimensions grow <u>exponentially</u> with r. Less obviously, the number of zero entries grows exponentially in r as well, although this effect may not be felt until relatively large r are encountered ($r > 7$, say). Thus sparsity is not likely to be significant for small r even if $\underset{\sim}{S}$ is relatively sparse (note $\underset{\sim}{R}$ and $\underset{\sim}{T}$ will generally be full). However, the product structure of the matrices does lend itself to the efficient formulation of algorithms for matrix-vector multiplication requiring only the storage of $\underset{\sim}{R}$, $\underset{\sim}{S}$, and $\underset{\sim}{T}$. Moreover one may expect that parallel or vector computer architectures would be highly advantageous in the implementation of such algorithms.

This suggests the use of matrix eigenvalue algorithms designed for use on sparse symmetric matrices with irregular zero-nonzero structure that require only knowledge of how the coefficient matrices act in products with vectors. For example, the Lanczos method is well suited to this [7]. One may consider as well Rayleigh quotient iteration in conjunction with an iterative linear system solver such as either a Lanczos-type method again [14] or a conjugate gradient method [15].

We have restricted our considerations to the computation of lower bounds to the eigenvalues of the total Hamiltonian, H. Actually, the only eigenvalues of physical interest are those possessing eigenvectors that have symmetry properties conforming with the Pauli exclusion principle. Russell [16] has made progress in the difficult post-processing problem of distinguishing physical from non-physical eigenvalues for large atoms. As yet, it is unclear how one might include such information in the initial formulation and to what extent the structure of the final matrix eigenvalue problem would suffer from such inclusions.

Freund [8] has successfully demonstrated a different computational approach in attempting to compute lower bounds to bound states of a Hamiltonian modeling a lithium atom. His work emphasized strategies in evaluating the complicated multi-dimensional integrals involved - a problem that certainly persists in our situation. By contrast, we have chosen to emphasize the linear algebraic aspects of the problem which are likely to dominate for atoms with somewhat larger atomic numbers.

§7. Acknowledgements. It is a pleasure to thank Professor Robert Nyden Hill for an enlightening discussion on aspects of the last section.

References

[1] Aronszajn, N.: Approximation Methods for Eigenvalues of Completely Continuous Symmetric Operators. Proc. Symp. Spectral Theory and Differential Problems, Stillwater (1951), 179-202.

[2] Bazley, N. W.: Lower Bounds for Eigenvalues with Applications to the Helium Atom. Phys. Rev. 120 (1960), 144-149.

[3] Bazley, N. W. and Fox. F. W.: Truncations in the Method of Intermediate Problems for Lower Bounds to Eigenvalues. Jour. Res. Nat. Bur. Stds. 65B (2) (1961), 105-111.

[4] Beattie, C.: Some Convergence Results for Intermediate Problems that Displace Essential Spectra. M.S.E. Research Center preprint series, J.H.U. Applied Physics Laboratory. Laurel, Maryland, No. 65 (1982).

[5] Beattie, C.: Convergence of Essential Spectra in Fox's Construction of Intermediate Hamiltonians. University of Arizona preprint (1983).

[6] Brown, R. D.: Variational Approximation Methods for Eigenvalues. Convergence Theorems. Proc. Banach Ctr., Warsaw, to appear.

[7] Cullum, J. and Willoughby, R.: Lanczos and the Computation in Specified Intervals of the Spectrum of Large, Sparse Real Symmetric Matrices. In Sparse Matrix Proceedings 1978, Philadelphia, SIAM (1979).

[8] Freund, D.: Upper and Lower Bounds to Two and Three Electron Systems. Dissertation, University of Delaware. Newark (1982).

[9] Fox, D. W.: Lower Bounds for Eigenvalues with Displacement of Essential Spectra. SIAM Jour. Math. Anal. 3 (4) (1972), 617-624.

[10] Fox, D. W. and Rheinboldt, W. C.: Computational Methods for Determining Lower Bounds for Eigenvalues of Operators in Hilbert Space. SIAM Rev. 8 (4) (1966) 427-462.

[11] Greenlee, W. M.: A Convergent Variational Method of Eigenvalue Approximation. Arch. Rat. Mech. Anal. 81 (3) (1983) 279-287.

[12] Greenlee, W. M. and Russell, D.: Progress on Estimation of Energy Levels for Multielectron Atoms, these proceedings.

[13] Klahn, B. and Bingel, W.: The Convergence of the Rayleigh-Ritz Method in Quantum Chemistry II. Theor. Chim. ACTA 44 (1977) 27-43.

[14] Paige, C. C. and Saunders, M. A.: Solution of Sparse Indefinite Systems of Equations. SIAM J. Numer. Anal. 12 (1975) 617-629.

[15] Reid, J. K.: The Use of Conjugate Gradients for Systems of Linear Equations Possessing "Property A". SIAM J. Numer. Anal. 9 (1972) 325-332.

[16] Russell, D. M.: The Calculation of Lower Bounds to Atomic Energies. Dissertation, University of Arizona, Tucson (1983).

[17] Thirring, W.: A Course in Mathematical Physics 3: Quantum Mechanics of Atoms and Molecules. New York, Springer (1981).

[18] Weinstein, A. and Stenger, W.: Methods for Intermediate Problems for Eigenvalues: Theory and Ramifications. New York Academic Press (1972).

Department of Mathematics
University of Arizona
Tucson, Arizona 85721
USA

ISNM, Vol. 69
Numerical Treatment of
Eigenvalue Problems, Vol. 3

HARTREE-FOCK METHODS
A REALIZATION OF VARIATIONAL METHODS IN COMPUTING ENERGY LEVELS IN ATOMS

Klaus Böhmer and Wolfgang Gross

In this paper the well known Hartree-Fock methods are interpreted as variational methods. Since good and reliable upper bounds for the lowest eigenvalue of the Schrödinger equation are very important, we discuss the different kinds of numerical errors during the computation and give some hints how to control them.

1. Introduction

After introducing the time-dependent and the time-independent Schrödinger equations we discuss some of the mathematical and physical problems in this context, define the variational problem and the corresponding Hartree-Fock approximation in § 2,3. In § 4 we report on the Hartree-Fock equations in integrodifferential and in differential equation form for the single configuration Hartree-Fock approach. § 5 is concerned with numerical problems occuring in solving the Hartree-Fock equations. Especially we want to accelerate the computations and describe tools for measuring their reliability. This is an important project since in many computer centers up to 80% of the computation time is spent on Hartree-Fock calculations and since no error estimates seem to be available.

2. Many Electron Systems

In quantum mechanics the state of a system of N different particles moving in a potential V is described by the time-dependent Schrödinger equation (see e.g. [8])

$$(1)\quad i\frac{\partial}{\partial t}\,\phi(\vec{r}_1,\ldots,\vec{r}_N,t) = (-\frac{1}{2}\sum_{k=1}^{N}\Delta_k + V(\vec{r}_1,\ldots,\vec{r}_N))\phi(\vec{r}_1,\ldots,\vec{r}_N,t)$$

In (1), Δ_k represents the Laplacian applied to $\vec{r}_k$, the position vector of particle k. $-\frac{1}{2}\Sigma\Delta_k$ describes the kinetic, V the potential energy of the system, respectively. We have assumed that no external forces are active.

Solutions ϕ have to satisfy the following conditions:

(2i) ϕ is twice differentiable in $(\vec{r}_1,\ldots,\vec{r}_N)$, once differentiable in t,

(2ii) $\phi \in L_2(\mathbb{R}^{3N})$ for t fixed,

(2iii) $\int_{\mathbb{R}^{3N}} |\phi(\vec{r}_1,\ldots,\vec{r}_N,t)|^2\, d\vec{r}_1\cdots d\vec{r}_N = 1$

For every function ϕ satisfying (2), the right and left hand side in (1) are well defined, especially the right hand side is called Hamiltonian, the ϕ are called wavefunctions. Usually the properties in (2) with respect to t are dropped if only the Hamiltonian is studied.

An ansatz for a solution of the form

$$\phi(\vec{r}_1,\ldots,\vec{r}_N,t) = \varphi(\vec{r}_1,\ldots,\vec{r}_N)\cdot e^{-iEt}$$

leads to the time - independent Schrödinger equation

$$(3)\qquad (-\frac{1}{2}\sum_{k=1}^{N}\Delta_k + V(\vec{r}_1,\ldots,\vec{r}_N))\varphi(\vec{r}_1,\ldots,\vec{r}_N) = E\,\varphi(\vec{r}_1,\ldots,\vec{r}_N)$$

which is an eigenvalue problem. Only solutions φ in $L_2(\mathbb{R}^{3N})$ are of interest.

The time independet Schrödinger equation for a many electron system is defined by its Hamiltonian.

The differential form for the non relativistic Hamiltonian is in this case given by

$$(4)\qquad \tau = -\frac{1}{2}\sum_{k=1}^{N}(\Delta_k + \frac{2Z}{r_k}) + \sum_{i<k}\frac{1}{r_{ik}}$$

where $r_{ik} = \|\vec{r}_i - \vec{r}_k\|$ is the distance between electron i and k, $r_i = \|\vec{r}_i\|$ the distance between electron i and the nucleus, N is the number of electrons, Z the charge of the nucleus. $1/r_{ik}$ represents the interaction potential between electrons i and k,

$-Z/r_k$ the potential between electron k and the nucleus, where we have used atomic units. In (4) we assume that the nucleus can be treated as a point charge with infinite mass.
The operator

$$H_0 : \begin{cases} C_0^\infty(\mathbb{R}^{3N}) \longrightarrow L_2(\mathbb{R}^{3N}) \\ \varphi \longrightarrow H_0\varphi = \tau\varphi \end{cases}$$

is essentially selfadjoint on $C_0^\infty(\mathbb{R}^{3N})$. Its closure H, called Hamiltonian, is selfadjoint, bounded from below and its essential spectrum is $\sigma_e = [\varkappa,\infty)$. There exist eigenvalues of finite multiplicity below $\varkappa$, the corresponding eigenfunctions are called bound states. (cf. [10,ch.10], [7,II.3,IV.4], [13,ch.10]).
A naive numerical approach to solve (3),(4) would be to replace $\Sigma\Delta_k$ by the usual symmetric difference formulae of order two, incorporating zero Dirichlet boundary conditions.
Now, consider the tabulation of one stationary state of Fe (Iron N=26), i.e. a solution of (3),(4). If only 10 discrete points are used for each of the 3N variables, this would amount to

$$10^{3N} = 10^{78}$$

unknowns. Using symmetry properties of the solutions we end up with

$$5^{78} \cong 10^{53} \quad ,$$

unknowns, so still far too many quantities to handle the situation. Hence, we need totally different numerical procedures.
These are based on the variational principle

> A pair (E,ϕ_E) is a solution of (3) iff ϕ_E is a stationary point of the Rayleigh quotient
>
> $$R_H(\phi) := \frac{(H\phi,\phi)}{(\phi,\phi)} \quad ,$$
>
> and $R_H[\phi_E] = E$.

Here $(H\phi,\phi)$, (ϕ,ϕ) denote the usual inner products in $L^2(\mathbb{R}^{3N})$. In this paper we concentrate on the lowest eigenvalue, E^0, which corresponds to the ground state of the atom. By the min-max-principle (cf. [9]) we have in the case of the above Hamiltonian

(5) $E^{0} = \min\{R_H(\phi) : \phi \in D(H)\}$,
and can use a Rayleigh-Ritz technique.

3. The approximating subsets

The Hartree-Fock methods are essentially variational methods applied to the problem (3). It is well known that in these methods good results depend decisively upon the choice of the subspaces in which (5) is solved approximately. These spaces should in fact reflect the physical nature of the problem as well as possible. This is not yet reflected well by equation (3). The condition that in a many electron system only "different particles" are possible is not satisfied in the case of atoms if only space variables are involved. One has to use spin coordinates to distinguish the particles. Further there is an effect of the spin on the energy called "spin-orbit-interaction" which is neglected in (4), so the wavefunctions separate into space part and spin part. The Hamiltonian (4) is symmetric with respect to an interchange of the 4 coordinates (3 space, 1 spin) of any two electrons. This means that any solution of (3) is a linear combination of symmetric and antisymmetric solutions. However, the only states observed are antisymmetric with respect to an interchange of all pairs of electrons. Hence, any good solution in a physical sense must satisfy the (see [4])

(A) Antisymmetry conditions.

If one could define an "average potential" $V(\cdot)$ such that

$$(6)\qquad \begin{aligned} H &= -\frac{1}{2}\sum_{k=1}^{N}\left(\Delta_k + \frac{2Z}{r_k}\right) + \sum_{i<k}\frac{1}{r_{ik}} \\ &\cong -\frac{1}{2}\sum_{k=1}^{N}\left(\Delta_k + \frac{2Z}{r_k} + V(r_k)\right) \end{aligned}$$

with $\cong$ not specified in this context, this would yield a product ansatz

$$(7)\qquad \begin{aligned} \varphi(\vec{r}_1,\sigma_1,\ldots,\vec{r}_N,\sigma_N) &\cong \phi(\vec{r}_1,\sigma_1,\ldots,\vec{r}_N,\sigma_N) \\ &= \varphi_1(\vec{r}_1,\sigma_1)\cdot\ldots\cdot\varphi_N(\vec{r}_N,\sigma_N) \end{aligned}$$

with an equation for every single φ_k as in the case of a central field, so

$$(8) \qquad -\frac{1}{2}\left(\Delta_k + \frac{2Z}{r_k} + V(r_k) \right) \varphi_k = E_k \varphi_k \qquad k=1,\ldots,N \quad .$$

It is well known from mathematical physics that these equations are solved by

$$(9) \qquad \varphi_k(\vec{r},\sigma) = \frac{1}{r} P(n,l,\ldots;r)\cdot Y_{l,m_l}(\vartheta,\varphi)\cdot \chi_{m_s}(\sigma)$$

where $\vec{r}$ is represented by its polar coordinates (r,ϑ,φ) and where the quantum numbers n,l,m_l,m_s, corresponding to k, give a more fundamental description of the electron than k.
$P(n,l,\ldots;r)$ depends, but the spherical harmonic Y_{l,m_l} does not depend upon Z and $V(\cdot)$. Therefore, instead of computing $V(\cdot)$ and determining $P(n,l,\ldots;r)$ via (6)-(9), we might as well start to compute the $P(n,l,\ldots;r)$ from appropriate conditions and obtain the approximation to φ in a form similar to (7).
Before we proceed to do so, we want to discuss the advantages and disadvantages of approximations of the form (9). For not too big atoms the interaction between different electrons will be small enough as to approximate the dependency upon the angular variables (ϑ,φ) well enough by the spherical harmonics $Y_{l,m_l}(\vartheta,\varphi)$ if only the $P(n,l,\ldots;r)$ are appropriately defined. By choosing (9) we certainly have fixed a certain angular behaviour of the wave-functions which does not allow to properly describe other angular behaviour. But since we have to choose some type of subspaces we do it this way.
Now let $\varphi_k(i)$ denote the "spin-orbital", that is defined by the quantum numbers (n,l,m_l,m_s) corresponding to k and evaluated for the coordinates of electron i. Then we assume φ to be a linear combination of products of "spin-orbitals"

$$(10) \qquad \varphi = \sum_{\alpha\in A} c(\alpha)\, \varphi_{\alpha(1)}(1) \cdot\cdots\cdot \varphi_{\alpha(N)}(N) \; .$$

with a given set A of N quantum numbers (n,l,m_l,m_s).

The Slater determinant

$$\text{(SD)} \qquad \varphi = \frac{1}{\sqrt{N!}} \begin{vmatrix} \varphi_1(1) & \cdots & \varphi_1(N) \\ \varphi_2(1) & & \varphi_2(N) \\ \vdots & & \vdots \\ \varphi_N(1) & \cdots & \varphi_N(N) \end{vmatrix}$$

is a linear combination of this form which in addition satisfies the antisymmetry condition.
Furthermore we usually require that the spin orbitals form an orthonormal set

$$\text{(O)} \qquad (\varphi_i, \varphi_k) = \delta_{ik} .$$

For the computation of a given state of an atom one knows the quantum numbers $\{n, l, m_l, m_s\}$ corresponding to this system, e.g. we have

q_1 electrons corresponding to $n_1 l_1$,
q_2 electrons corresponding to $n_2 l_2$,
.
.
.
q_m electrons corresponding to $n_m l_m$,

usually indicated as

$$(n_1 l_1)^{q_1} (n_2 l_2)^{q_2} \cdots\cdots (n_m l_m)^{q_m}$$

for an atom with

$$N = q_1 + q_2 + \dots + q_m$$

electrons. Since the spherical harmonics $Y_{l,m_l}(\vartheta, \varphi)$ are known, the problem is reduced to the computation of the radial functions $P(n, l, \dots; r)$. Since further the $Y_{l,m}$ have the well known orthogonality properties, (O) has to be required only for radial functions.
So we have the following result: Instead of (5) solve the following variational problem:

$$(10) \qquad E^{0}_{HF} = \min \{ R_H(\phi) : \phi \in S_{HF} \} .$$

The set S_{HF} is defined for a given set A of N quantum numbers

(11) $A = \{ (n_1 l_1 m_1^1),\dots,(n_1 l_1 m_{q_1}^1),\dots,(n_m l_m m_1^m),\dots,(n_m l_m m_{q_m}^m) \}$

abbreviated into

(12) $\{1,2,\dots,N\}$

and corresponding to the state

$$(n_1 l_1)^{q_1} \cdots (n_m l_m)^{q_m} \quad .$$

S_{HF} is a subset of $L_2(\mathbb{R}^{3N})$ containing every Slater determinant determinant (SD) with φ_k of the form (9) corresponding to the quantum numbers (11),(12) with $P(n,l,\dots;r) \in L_2(\mathbb{R})$ to be determined such that (O) is satisfied.

Here we only study the socalled Single Configuration Hartree-Fock method, that is, only one Slater determinant is used for the approximation, in contrast, linear combinations of a finite number of Slater determinants are used in Multi-Configuration Hartree-Fock methods.

4. The Hartree-Fock Equations

Having defined S_{HF} as a subset of $L_2(\mathbb{R}^{3N})$, the well known results on variational methods immediately yield

$$E^0 \leq E_{HF}^0 \quad .$$

Now, if we set up the necessary conditions to solve (10) in S_{HF} with φ_k of the form (9) we end up with the following system of integrodifferential equations for the radial parts of the wave-functions

$$\mathbf{L}_i \equiv -\frac{d^2}{dr^2} - \frac{2Z}{r} + \frac{l_i(l_i+1)}{r^2}$$

$$
(\mathrm{HFI})\quad
\begin{cases}
L_i p_i(r) + \frac{2}{r} \sum_{j=1}^{m} \int_0^\infty g_{ij}(r,s) p_j^2(s)ds \cdot p_i(r) + \\
\quad + \frac{2}{r} \sum_{\substack{j=1 \\ j\neq i}}^{m} \int_0^\infty h_{ij}(r,s) p_i(s) p_j(s) ds \cdot p_j(r) + \\
\quad + \sum_{j} \varepsilon_{ij}\, p_j(r) = 0 \quad , \quad i=1(1)m \quad , \\
(p_i, p_j) = \delta_{ij} \quad (i,j)\in \mathbf{I} \quad \text{orthogonality constraints,} \\
p_i(0) = \lim_{r\to\infty} p_i(r) = 0 \quad \text{boundary conditions.}
\end{cases}
$$

Defining

$$
U^k(r,s) = \begin{cases} \left(\frac{s}{r}\right)^k & s \le r \ , \\ \left(\frac{r}{s}\right)^{k+1} & r < s \ , \end{cases}
$$

the g_{ij} and h_{ij} are given by

$$
g_{ij}(r,s) = \sum_{k=0}^{\cdots} A_{ijk} U^k(r,s) \quad ,
$$

$$
h_{ij}(r,s) = \sum_{k=0}^{\cdots} B_{ijk} U^k(r,s) \quad .
$$

Here $m, Z, l_i, A_{ijk}, B_{ijk}$, $\mathbf{I} \subset \{1,\dots,m\}\times\{1,\dots,m\}$ depend on the configuration of the atom and are prescribed values. (The computation of A_{ijk}, B_{ijk} is a rather lengthy problem.)

The first term $\frac{2}{r}\sum \int g_{ij}(r,s)p_j^2(s)ds$ represents the screening potential observed by electron i due to the other electrons, the term $\frac{2}{r}\sum \int h_{ij}(r,s)p_i(s)p_j(s)ds$ is called exchange potential and reflects Pauli's principle.

(HFI) is a coupled system of nonlinear integrodifferential equations on an infinite intervall with boundary conditions and orthogonality constraints. It is an eigenvalue problem.

We have gained very much already: For iron (Fe,N=26) the system (1),(3) needed 3N=78 variables to describe the situation and to compute the 10^{78} or 10^{53} unknowns. We need for the groundstate of iron,

$$1s^2 2s^2 2p^6 3s^2 3p^6 3d^6 4s^2 \quad ,$$

only 7 groups of electrons, therefore m=7 in (HFI), or, for a similar accuracy 7·10 unknowns.

Since the solution of the full matrices corresponding to any discretization of (HFI) requires much computational time, we prefere to transform (HFI) into a system of differential equations, even if more functions should be required. This is in effect possible, since the functions

$$\Gamma^k(r,s) := \frac{s}{2k+1} U^k(r,s)$$

are Green's functions for the second order problem

$$\mathbf{M}_k z \equiv -z'' + \frac{k(k+1)}{r^2} z = 0 \qquad z(0)=0,\ z(\infty)=0.$$

Hence the (HFI) equations are equivalent to the following system

$$(\mathrm{HFD}) \left\{ \begin{array}{ll} \mathbf{L}_i p_i + \frac{2}{r} \sum_{j,k} a_{ijk}\, y_{jjk}\, p_i + \frac{2}{r} \sum_{\substack{j,k \\ j\neq i}} b_{ijk}\, y_{ijk}\, p_j + & \\ \qquad + \sum_j \varepsilon_{ij}\, p_j = 0 & i=1,\ldots m \\ \mathbf{M}_k\, y_{ijk} - \frac{2k+1}{r} p_i p_j = 0 & \begin{array}{l} i,j=1,\ldots,m \\ k\in I(i,j) \end{array} \\ (p_i, p_j) = \delta_{ij} \qquad (i,j)\in \mathbf{I} \quad , & p_i(0)=\lim_{r\to\infty} p_i(r)=0 \\ y_{ijk}(0)=\lim_{r\to\infty} y_{ijk}(r)=0 & \end{array} \right.$$

Here the integral terms are removed to yield a coupled system of differential equations. In general only few of the coefficients a_{ijk} and b_{ijk} are not equal to zero. Hence only the corresponding y_{ijk} need to be computed.

5. Numerical Methods for (HFD)

From physical arguments one knows very well the asymptotic behaviour of the $p_i(r)$ for $r\to 0$. The $p_i(r)$ tend to zero for $t\to\infty$, and this happens already for not to large r. Therefore one tranforms the intervall $(0,\infty)$ via

$$\rho := \log r$$

into $(-\infty,\infty)$ and uses an equidistant grid. The differential equations (HFD) are transformed as follows:

$$\bar{L}_i z(\rho) \equiv \left(-\frac{d^2}{d\rho^2} - 2Ze^{\rho} + (l_i+\tfrac{1}{2})^2 \right) z(\rho) ,$$

$$\bar{M}_k z(\rho) \equiv \left(-\frac{d^2}{d\rho^2} + (k+\tfrac{1}{2})^2 \right) z(\rho) ,$$

$$\bar{p}_i(\rho) := \frac{p_i(r)}{\sqrt{r}} , \quad \bar{y}_{ijk}(\rho) := \frac{y_{ijk}(r)}{\sqrt{r}} ,$$

$$(\overline{HFD}) \left\{ \begin{array}{ll} \bar{L}_i\bar{p}_i + 2e^{3\rho/2}\left(\sum\limits_{j,k} a_{ijk}\,\bar{y}_{jjk}\,\bar{p}_i + \sum\limits_{\substack{j,k \\ j\neq i}} b_{ijk}\,\bar{y}_{ijk}\,\bar{p}_j \right) + & \\ \quad + \; e^{2\rho} \sum\limits_{j} \varepsilon_{ij}\,\bar{p}_j = 0 & i=1,\dots m \\ M_k\,\bar{y}_{ijk} - (2k+1)\,e^{3\rho/2}\,\bar{p}_i\bar{p}_j = 0 & i,j=1,\dots,m \quad k\in I(i,j) \\ (\bar{p}_i,\bar{p}_j) = \delta_{ij} \quad (i,j)\in \mathbf{I} \; , \quad \lim\limits_{\rho\to\pm\infty} \bar{p}_i(\rho)=0 & \\ \lim\limits_{\rho\to\pm\infty} \bar{y}_{ijk}(\rho)=0 & \end{array} \right.$$

Let $(\overline{HFD})^h_{(-\infty,\infty)}$ denote the discretization of $(\overline{HFD})_{(-\infty,\infty)}$ on an equidistant grid either by the classical second order symmetric difference method or by the Collatz Mehrstellenverfahren [3], and $(\overline{HFD})^h_{(a,b)}$ the corresponding discretization, where in $(\overline{HFD})_{(-\infty,\infty)}$ the $-\infty,\infty$ are replaced by a,b , respectively. This transform has been used before.

Ch. Froese-Fischer has described the Hartree-Fock methods in [4] and implemented a numerical method for (HFD) in [5], [6]. She uses essentially multiple shooting combined with Collatz Mehrstellenverfahren and applies deferred corrections, to obtain higher order methods. The discrete system of difference equations, obtained from (HFD) by this process, is then solved by a block successive iteration method. With starting values for the

p_i , $i=1,...,m$, in grid points, discrete approximations for the y_{ijk} , and then for the p_i are computed from discretizations of the operators $\mathbf{M}y_{ijk}=...$, $\mathbf{L}_i p_i=...$ and boundary conditions, followed by an orthonormalization step. Problems arise from slow convergence and from the fact that an estimation for the errors is hardly possible because of the complicated agglomeration of the numerical procedures. Further, the deferred corrections applied to the Collatz Mehrstellenverfahren do not account for the fact that the infinite intervals in $(\overline{HDF})_{(-\infty,\infty)}$ are simply reduced to finite subintervals $(\overline{HFD})_{(a,b)}$ without discussing the corresponding error.

Our group at Marburg has chosen the following approach. Asymptotic expansions for the truncation error from $(\overline{HFD})_{(-\infty,\infty)}$ to $(\overline{HFD})_{(a,b)}$ are obtained for the operators $\bar{\mathbf{L}}_i$ and $\bar{\mathbf{M}}_k$ by Schwarz in [12]. The equations for the y_{ijk} have a simple structure, but these are roughly $m^3/3$ equations. Hence it is desirable to solve them with maximum efficiency. Using the above difference scheme and a special form of the boundary conditions one obtains a factorization of the discretization matrix

$$\begin{pmatrix} \mu & -1 & & & \\ -1 & c & -1 & & \\ & \ddots & \ddots & \ddots & \\ & & -1 & c & -1 \\ & & & -1 & \mu \end{pmatrix} = \begin{pmatrix} 1 & & & & \\ -\sigma & 1 & & & \\ & \ddots & \ddots & & \\ & & -\sigma & 1 & \\ & & & -\sigma & 1 \end{pmatrix} \begin{pmatrix} \mu & -1 & & & \\ & \mu & -1 & & \\ & & \ddots & \ddots & \\ & & & \mu & -1 \\ & & & & \mu-\sigma \end{pmatrix}$$

$$c>2 \qquad \mu^2-c\mu+1=0 \qquad \sigma=1/\mu$$

This leads to a reduction of the number of operations by a factor of 0.6 compared to the standard implementation, see Schmitt in [11].

Discrete Newton methods, see Böhmer [2], are used with these results to obtain reliable error estimates for the truncation errors - from $(\overline{HFD})_{(-\infty,\infty)}$ to $(\overline{HFD})_{(a,b)}$ - and the discretization errors - from $(\overline{HFD})_{(a,b)}$ to $(\overline{HFD})^h_{(a,b)}$.

The nonlinear system $(\overline{HFD})^h_{(a,b)}$ is solved by applying the Newton method combined with an extensive use of the special structure of the Jacobian. Apparently this has not been implemented on a large scale so far. The structure of the Jacobian in case all k=0 in $(\overline{HFD})^{h}_{(a,b)}$ is as follows. Let

$$\mathbf{X} = (p_1,\dots,p_m,y_{110},y_{120},\dots,y_{1m0},y_{220},\dots\dots,y_{mm0})$$

neglecting the λ_{ij}, boundary conditions etc which give an additional bordering of the matrix. Then the Jacobian has the form

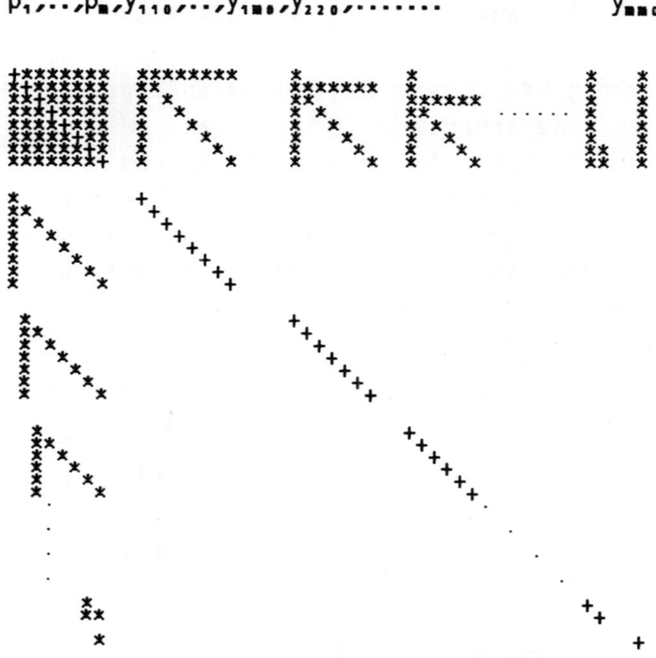

Each entry * is a diagonal matrix. The entries + in the main diagonal are tridiagonal matrices, corresponding to the discretization. Note that $y_{ijk} = y_{jik}$. The headline indicates the correspondence of the +,* to the unknown $p_1,\dots,y_{mm0}$.

A lot is known analytically about the radial functions - for instance the number of zeros - and can be used extensively during the computation. Especially this information is very useful to start with a very coarse step size h_0, use Newton's method on $(\overline{HFD})^{h_0}_{(a,b)}$ and proceed to the necessary small step sizes h in

$(\overline{HFD})^h_{(a,b)}$ using the mesh independence results in Allgower-Böhmer [1] and the corresponding procedures, again reducing the computation time and yielding reliable error estimates.
We are now in the process to further develop all these tools and then to combine them to get good and reliable error estimates and to reduce the computation time.

Literatur

[1] ALLGOWER,E;BÖHMER,K.: A mesh independence principle for operator equations and their discretizations. Sonderforschungsbereich 72, Universität Bonn, preprint 456

[2] BÖHMER,K.: Discrete Newton methods and iterated defect corrections. Num.Math.37,(1981), 167-192

[3] COLLATZ,L.: The numerical treatment of differential equations. 3.ed.-Berlin-Heidelberg-New York, Springer 1966

[4] FROESE-FISCHER,CH.: The Hartree-Fock method for atoms. New York-London-Sydney-Toronto, Wiley 1977

[5] FROESE-FISCHER,CH.: MCHF 72.-Comp.Phys.Comm.Journal 4 (1972) 107-

[6] FROESE-FISCHER,CH.: A general multi-configuration Hartree-Fock program. Comp. Phys. Comm. Journal 14 (1978),145-153

[7] HELLWIG,G.: Differentialoperatoren der mathematischen Physik Berlin-Göttingen-Heidelberg, Springer 1964

[8] PRUGOVECKI,E.: Quantum mechanics in Hilbert Space. 2.ed. New York-London-Toronto-Sydney-San Francisco,Academic Press 1981

[9] REED,M.; SIMON,B.: Methods of modern mathematical physics. Analysis of operators. New York-San Francisco-London, Academic Press 1978

[10]SCHECHTER,M.: Spectra of partial differential operators. Amsterdam-London, North Holland 1971

[11]SCHMITT,B.: Defect corrections on infinite intervals. To appear

[12]SCHWARZ,R.: Asymptotic boundary conditions for boundary value problems on an infinite interval. To appear.

[13]WEIDMANN,J.: Lineare Operatoren in Hilberträumen. Stuttgart, Teubner 1976

Klaus Böhmer
Wolfgang Gross

Fachbereich Mathematik
Philipps-Universität Marburg
Lahnberge
3550 Marburg/L.

ISNM, Vol. 69
Numerical Treatment of
Eigenvalue Problems, Vol. 3

SPECTRAL ANALYSIS OF NON-SELFADJOINT OPERATORS OF MATHEMATICAL PHYSICS

Jozef Brilla

The application of Laplace transformation to a class of non-stationary problems of mathematical physics leads to analysis of non-selfadjoint regular elliptic operators. It is shown that these operators have in the right-hand side half-plane of the transform parameter discrete regular spectrum and eigenelements form a complete set biorthonormal with eigenelements of the adjoint operator which are complex conjugate. Finally finite element solution is proposed.

1. INTRODUCTION

Different non-stationary problems of mathematical physics can be formulated as initial-boundary value problems of the following class of partial differential equations

$$(1) \qquad \sum_{k=0}^{n} A_k \frac{\partial^k}{\partial t^k} u = f \, , \qquad \text{in } \Omega \, ,$$

where A_k are symmetric elliptic operators of the same order, with necessary initial and boundary conditions. This class of equations includes parabolic, hyperbolic and quasiparabolic and quasihyperbolic equations. Quasiparabolic and quasihyperbolic equations are common in the theory of viscoelastic bodies and structures [2].

I.C. Gochberg and M.G. Krein [5] deal with the theory of non-selfadjoint operators of the form (1) assuming that A_k are compact operators. In the case of the homogeneous equation (1) using the substitution $u = \varphi \exp(\lambda t)$ they arrive at

$$(2) \qquad \sum_{k=0}^{n} \lambda^k A_k \varphi = 0 .$$

Then values λ, for which a non-trivial solution of (2) exists are called eigenvalues and corresponding non-trivial solutions eigenfunctions. Thus (2) represents a nonlinear eigenvalue problem, which leads to difficulties in connection with completness of the set of eigenfunctions. We have applied [3,5] to an analysis of (1) the Laplace transformation and arrived at

$$(3) \qquad \sum_{k=0}^{n} p^k A_k \tilde{u} = \tilde{f}^*,$$

where p is the parameter of the Laplace transformation, a tilde denotes a Laplace transform and $\tilde{f}^*$ includes also initial conditions. Then we define an eigenvalue and eigenfunction problem in the classical way and arrive at

$$(4) \qquad A\varphi = \sum_{k=0}^{n} p^k A_k \varphi = \lambda \varphi ,$$

where λ and φ are then functions of the transform parameter p. The operator A is a non-selfadjoint regular elliptic operator.

2. ANALYSIS OF THE OPERATOR A

In order to have a convenient basis for an analysis of non-selfadjoint regular elliptic operators we introduce complex Sobolev spaces $H^m(\Omega, p_\sigma^+)$ of functions, which are paparametrically dependent on the transform parameter p and analytic in the right-hand side half-plane $p_\sigma^+ = \{p \mid \operatorname{Re} p \geq \sigma \geq 0\}$ with the scalar product

$$(\tilde{u}, \tilde{v}) = \int_\Omega \sum_{|\alpha| \leq m} D^\alpha \tilde{u} \, D^\alpha \bar{\tilde{v}} \, d\Omega \tag{5}$$

and the corresponding norm, α beeing a multiindex.

Now we restrict ourselves to an analysis of the equation

$$A(p)\,\tilde{u} = \sum_{k=0}^{2} p^k A_k \tilde{u} = \tilde{f}^* \quad \text{in } \Omega, \tag{6}$$

with Dirichlet boundary conditions, where Ω is a bounded domain in R^n with a smooth boundary $\partial\Omega$.

We define the bilinear form

$$H[\tilde{u}, \tilde{v}] = \frac{1}{(p+\alpha)(p+\beta)} \sum_{k=0}^{2} p^k (A_k\tilde{u}, \tilde{v}) \tag{7}$$

for $\tilde{u}, \hat{v} \in H_0^m(\Omega, p_6^+)$ and $\alpha > 0$, $\beta > 0$.

$H[\tilde{u}, \tilde{u}]$ can assume zero values only at

$$p_{1,2} = -\frac{(A_1\tilde{u}, \tilde{u})}{2(A_2\tilde{u}, \tilde{u})} [1 \pm (1 - \frac{4(A_0\tilde{u}, \tilde{u})(A_2\tilde{u}, \tilde{u})}{(A_1\tilde{u}, \tilde{u})^2})^{1/2}] . \tag{8}$$

Since A_k are symmetric elliptic operators zero values of $H[\tilde{u}, \tilde{u}]$ can occur only in the left-hand side of the halfplane of p. It is possible to prove that for $p \in p_6^+$ the bilinear form (7) fulfils the conditions of the Lax - Milgram theorem. Thus the operator $A(p)/(p+\alpha)(p+\beta)$ has an inverse for each $p \in p^+$ and this inverse is a compact operator.

Further we have

$$A(p) = B(p-p_o) = B_2(p-p_o)^2 + B_1(p-p_o) + B_0 , \tag{9}$$

where

$$B_2 = A_2, \; B_1 = 2A_2p_o + A_1, \; B_0 = A_2p_o^2 + A_1p_o + A_0. \tag{10}$$

Then

$$(11)\qquad B^{-1}(p-p_o) = B_0^{-1} - (p-p_o)\, B_0^{-1}B_1B_0^{-1} + (p-p_o)^2(B_0^{-1}B_1B_0^{-1}B_1B_0^{-1} - B_0^{-1}B_2B_0^{-1}) + \dots$$

As B_0^{-1} is a compact operator there exists $\varepsilon > 0$ such that for $|p-p_o| < \varepsilon$ $B^{-1}(p-p_o)$ is a compact operator for each $p_o \in P_6^+$.

3. SPECTRAL ANALYSIS OF $A^{-1}(p)$

Operators $A(p)$, when A_0, A_1 and A_2 are noncommutative, are nonnormal. We shall prove that they are normaloid. The adjoint operator to $A^*(p)$ is its complex conjugate operator. It holds

$$(12)\qquad A^*(p) = \overline{A(p)}.$$

The same holds for its inverse $K(p) = A^{-1}(p)$. We have

$$(13)\qquad (A^{-1}(p))^* = K^*(p) = \overline{K(p)}.$$

In agreement with the matrix theory we call compact operators with this property Hermitian. We shall show that many results on spectral theory of symmetric compact operators can be generalized for Hermitian compact operators.

Consider the bilinear form $(Kf,\bar{f})$. Then it is obvious that

$$(14)\qquad |(Kf,\bar{f})| \leq \|Kf\|\,\|\bar{f}\| \leq \|K\|\ \|f\|^2.$$

When we denote r_K the smallest constant, for which it holds

$$(15)\qquad |(Kf,f)| \leq r_K \|f\|^2,$$

then

$$(16)\qquad r_K \leq \|K\|.$$

Now we take into account that $(K^2f,\bar{f}) = (Kf,\overline{Kf})$. Then as it holds

$$(17) \qquad |(Kf,\overline{Kf})| = (Kf,Kf) = \|Kf\|^2,$$

we have for each $c > 0$.

$$(18) \qquad \|Kf\|^2 = \left|\frac{1}{4}[(K(cKf + \frac{1}{c}f), \overline{cKf + \frac{1}{c}f}) - (K(cKf - \frac{1}{c}f), \overline{cKf - \frac{1}{c}f})]\right|$$

$$\leq \frac{1}{4}[r_K \|cKf + \frac{1}{c}f\|^2 + r_K \|cKf - \frac{1}{c}f\|^2]$$

$$= \frac{1}{2} r_K(c^2 \|Kf\|^2 + \frac{1}{c^2} \|f\|^2).$$

When $\|Kf\| \neq 0$, the right hand side of (18) assumes minimum for $c^2 = \|f\|/\|Kf\|$. Hence

$$(19) \qquad \|Kf\|^2 \leq r_K\|Kf\| \; \|f\| \quad , \qquad \|Kf\| \leq r_K \|f\| ,$$

what gives $\|K\| \leq r_K$ and thus with respect to (16) $r_K = \|K\|$. r_K is the numerical and spectral radius of K.

Hence Hermitian compact operators are normaloid. Further we can prove that there exists an eigenvalue μ_1 for which

$$(20) \qquad |\mu_1| = \|K\|$$

and we can formulate

THEOREM 1. Each Hermitian compact operator has at leat one eigenvalue different from zero, for which it holds

$$(21) \qquad \max_{\|f\|=1} |(Kf,\bar{f})| = |\mu_1|$$

and one or a finite number of eigenfunctions φ_{1i}, which fulfil the equation

$$(22)\qquad K\varphi_{1i} = \mu_1 \varphi_{1i},$$

Now by a reasoning similar to ,that for symmetric compact operators we can prove that a Hermitian compact operator has a finite or a countable set of eigenvalues and eigenfunctions.

For the adjoint operator $\overline{K}$ it holds

$$(23)\qquad \overline{K}\,\overline{\varphi}_k = \overline{\mu}_k \overline{\varphi}_k.$$

Thus eigenvalues and eigenfunctions of the operator adjoint to a Hermitian compact operator are complex conjugate. It holds

$$(24)\qquad (K\varphi_k, \overline{\varphi}_l) = \mu_k(\varphi_k, \overline{\varphi}_l),$$

$$(K\varphi_l, \overline{\varphi}_k) = \mu_l(\varphi_l, \overline{\varphi}_k)$$

and then

$$(25)\qquad (\mu_k - \mu_l)(\varphi_k, \overline{\varphi}_l) = 0,$$

Thus φ_k and $\overline{\varphi}_l$ form biorthogonal or biorthonormal sets of eigenfunctions.

We can expand each function $f \in H^o(\Omega, p^+)$ for a fixed value of p in the form

$$(26)\qquad f = \sum_{k=1}^{\infty} (f, \overline{\varphi}_k)\,\varphi_k$$

and prove similarly as it was done for symmetric compact operators in [1]

THEOREM 2: A Hermitian compact operator K(p) has for a fixed $p \in p_6^+$ a countable set of eigenfunctions φ_k , which are biorthogonal to complex conjugate eigenfunctions of the adjoint operator $\overline{K(p)}$. This sets of eigenfunctions are complete. Thus it holds (26) and

(27) $$\|f\|_{H^0} = \sum_{k=1}^{\infty} |(f, \varphi_k)|^2 .$$

Similar results are valid also for the sets of eigenvalues and eigenfunctions of Hermitian elliptic operators A(p).

4. ANALYTICALLY SYMMETRIC OPERATORS

Operators A(p) (4) are analytic functions of the parameter p and are symmetric elliptic operators for real nonnegative values of p. We call such operators analytically symmetric elliptic operators.

When we rewrite A(p) in the form (9) it is obvious that $B(p-p_o)$ is for each real $p_o \geqslant 0$ a regular operator [7], which is for real nonnegative values of p sef-adjoint. Thus we can generalize the theorem of F. Rellich on a regular discrete spectrum and we have:

THEOREM 3: For each real $p \geqslant 0$ the operator A(p) has a regular discrete spectrum, it is there exist eigenvalues $\lambda_1(p)$, $\lambda_2(p)$, ..., and eigenfunctions $\varphi_1(p)$, $\varphi_2(p)$,... all regular and such that for real nonnegative values of p the following holds

(28) $$A(p)\, \varphi_k(p) = \lambda_k(p)\, \varphi_k(p),$$

$\lambda_1(p)$, $\lambda_2(p)$,... is a complete orthonormal system,

(29) $$\lim_{k \to \infty} |\lambda_k(p)| = \infty .$$

$\lambda_k(p)$ and $\varphi_k(p)$ for complex values of $p \in p_\delta^+$ can be obtained by the analytic continuation. φ_k and $\overline{\varphi}_k$ form complete biortogonal or biorthonormal systems of eigenfunctions.

The existence of a regular discrete spectrum for complex values of $p \in p_\delta^+$ and for analytically nonsymmetric regular operators can be proved taking into account the biorthonormality in the same way as it was done by F. Rellich [7] for real values of p. He considers only orthonormal systems of eigenfunctions, what restricts his consideration and an application of the Weierstrass preparation theorem only to real values of the parameter ε. We shall give the proof of the theorem for complex values of p in an another paper.

5. ZERO POINTS OF EIGENVALUES

The solution of the non-homogeneous differential equation

$$A(p)\,\tilde{u} = \tilde{f} \tag{30}$$

in the form of the expansion in the series of eigenfunctions assumes the form

$$\tilde{u} = \sum_{k=1}^{\infty} \frac{1}{\lambda_k(p)} (\tilde{f}, \bar{\varphi}_k)\, \varphi_k. \tag{31}$$

Thus for an analysis of the solution zero values of $\lambda_k(p)$ are impotrant. Points p_k at which $\lambda_k(p)$ assume zero values are nonlinear eigenvalues in the sense of I.C. Gochberg and M.G. Krein [5]. We can rewrite (31) in the form

$$\tilde{u} = \sum_{k=1}^{\infty} \frac{1}{(A_2\varphi_k, \bar{\varphi}_k)} \frac{1}{p_{k1} - p_{k2}} \left(\frac{1}{p - p_{k1}} - \frac{1}{p - p_{k2}} \right) (\tilde{f}, \bar{\varphi}_k)\, \varphi_k, \tag{32}$$

where

$$p_{k1,2} = - \frac{(A_1\varphi_k, \bar{\varphi}_k)}{(A_2\varphi_k, \bar{\varphi}_k)} \left[1 \pm \left(1 - \frac{4(A_0\varphi_k, \bar{\varphi}_k)(A_2\varphi_k, \bar{\varphi}_k)}{(A_1\varphi_k, \bar{\varphi}_k)^2}\right)^{1/2}\right]. \tag{33}$$

From the point of view of a convergence of the solution and approximate solutions and the global analysis limits of $p_{k1,2}$ for $k \to \infty$ are important. $p_{k1,2}$ are points for which $\lambda(p) = 0$ and then we have

$$(34) \qquad A(p_k)\,\varphi(p_k) = 0.$$

Thus it also holds

$$(35) \qquad (A(p_k)\,\varphi(p_k),\ \varphi(p_k)) = 0$$

and with respect to the analysis of (8) $p_{k1,2} \in p_{\sigma}^{+}$.

As

$$(36) \qquad \lim_{k\to\infty} (A_\alpha \varphi_k, \varphi_k) \to \infty \qquad \alpha = 0,1,2$$

and

$$(37) \qquad \lim_{k\to\infty} (A_\alpha \varphi_k, \varphi_k)/(A_\beta \varphi_k, \varphi_k) < \infty \quad , \alpha, \beta = 0,1,2,$$

it holds:

1. for $A_2 \neq I$

$$(38) \qquad \lim_{k\to\infty} |p_{k1,2}| < \infty$$

2. for $A_2 = I$

$$(39) \qquad \lim_{k\to\infty} p_{k1} \to -\infty \quad , \quad \lim_{k\to\infty} p_{k2} > -\infty \ .$$

This difference is important also from the point of view of functional spaces we have introduced [3,4] for an analysis of this class of equations.

6. APPROXIMATE SOLUTION

When $\tilde{f} \in H^m(\Omega, p_\sigma^+)$ then for a fixed $p \in p_\sigma^+$ Re $\tilde{f}$ and Im $\tilde{f} \in H^m(\Omega)$. Thus we can choose a basis $e_1, \ldots, e_n$ of a finite dimensional subspace of $H^m(\Omega)$ as a basis of the finite dimensional subspace of $H^m(\Omega, p_6^+)$. Hence we can seek for an approximate solution of the nonhomogeneous problem under consideration in the form

$$(40) \qquad \tilde{u}_n = \sum_{k=1}^{n} a_k(p) \; e_k(x) ,$$

where in contradistinction to a symmetric elliptic case independent of a parameter coefficients of the approximate solution are functions of the parameter p. When we are looking for eigenvalues, we expand the approximate value of an eigenfunction in the form

$$(41) \qquad \varphi_n = \sum_{k=1}^{n} a_k(p) \; e_k(x).$$

Applying the Galerkin method yields the equation for the determinationof $a_k(p)$ in the form

$$(42) \qquad \sum_{k=1}^{1} ([A(p) - \lambda(p)]e_k, \; e_l) \; a_k(p) = 0 .$$

Hence $\lambda(p)$ has to be a solution of the determinantal equation

$$(43) \qquad |([A(p) - \lambda(p)]e_k, \; e_l)| = 0.$$

Numerically we get different values of $\lambda(p)$ for different fixed values of p.

Zero values of $\lambda(p)$ are at those values of p, which fulfil the determinantal equation

$$(44) \qquad |(A(p)e_k, \; e_l)| = 0.$$

An error estimation in the n-th approximation for a fixed $p \in p_6^+$ can be obtained from the formula [6]

$$(45) \qquad |\lambda_k^{-1} - \lambda_{k,n}^{-1}| \leq \|K - P_n K P_n\| ,$$

where P_n is the corresponding projection operator and $K = A^{-1}$.

REFERENCES

1. Achieser,N.I., Glasmann, I.M.: Theorie der linearen Operatoren im Hilbert-Raum. Berlin, Akademie-Verlag 1968.
2. Brilla, J.: Generalized variational methods in linear viscoelasticity. In: Hult, J.(ed.). Mechanics of Viscoelastic Media and Bodies. Berlin-Heideberg-New York, Springer 1975, 215-228.
3. Brilla, J.: Error Analysis for Laplace Transform - Finite Element Solution of Hyperbolic Equations. Numer. Math. 41 (1983), 55-62.
4. Brilla, J.: New Functional Spaces and Analysis of Nonstationary Problems of Mathematical Physics. In: Greguš, M.(ed.). Equadiff V. Proc. of 5-th Czechoslovak Conf. on Diff. Eqns. and Their Appl. Leipzig, Teubner 1982, 64-71.
5. Gohberg, I.C., Krein, M.G.: Introduction to the Theory of Linear Nonselfadjoint Operators.Providence, Am. Math. Soc. 1969.
6. Krasnoselskii, M.A., Vainikko, G.M., Zabreiko, P.P., Rutitskii, Ya.B., Stetsenko, V.Ya.: Approximate Solution of Operator Equations. Groningen, Wolters-Noordhoff 1972.
7. Rellich, F.: Perturbation Theory of Eigenvalue Problems. New York-London-Paris, Gordon and Breach 1969.

Jozef Brilla
Comenius University
Institute of Applied Mathematics
and Computing Technique
84215 Bratislava
Czechoslovakia

ISNM, Vol. 69
Numerical Treatment of
Eigenvalue Problems, Vol. 3

An Inclusion Principle for Eigenvalues

Ulrich Eckhardt

Abstract. A general inclusion principle for eigenvalues with special properties (e.g. belonging to nonnegative eigenvectors) is developed and compared with Collatz's theorem.

In his famous paper [1] (see also [2]) L. Collatz proved inclusion theorems for eigenvalues of matrices. These theorems have the general form:

Given a real (n,n)-matrix A, one is looking for nontrivial solutions λ, x of the eigenvalue problem

$$Ax = \lambda \cdot x. \tag{1}$$

For a test vector $u = (u_1, \ldots, u_n)^T \in \mathbb{R}^n$ with nonzero components let $v = (v_1, \ldots, v_n)^T := Au$ and define numbers

$$\mu_i = v_i/u_i, \qquad i=1,\ldots, n. \tag{2}$$

Under the assumptions

(a) A is symmetric,

or

(b) A has only nonnegative elements and the components of u are positive,

Collatz showed that there is at least one eigenvalue λ of A with

$$m := \min \mu_i \leqq \lambda \leqq \max \mu_i =: M. \tag{3}$$

Theorems of this type proved as very useful tools in numerical analysis. Today, even on small computers, it is possible to investigate fairly large matrices by means of such simple inclusions. The knowledge of the mathematician or engineer about his problem influences directly the precision of the inclusion by using intelligent choices for test functions. A quick check on a peripheral intelligent terminal might save labour and computer time on the central large computer. In case of long responce times of the main computer system the previous inspection of the problem on an intelligent terminal can be very desirable.

Now let M_λ be a real (n,n)-matrix depending on a parameter λ. Suppose we are interested in solutions x of

$$M_\lambda x = \Theta \tag{4}$$

(Θ = zero vector in $\mathbb{R}^n$) under the additional requirement that

$$Cx \geqq \Theta, \qquad Cx \neq \Theta. \tag{5}$$

The (n,n)-matrix C is used to express desirable properties of the eigenvector belonging to the eigenvalue to be estimated. Such properties are: Nonnegativity, monotonicity, convexity, or a prescribed sign distribution.

<u>Lemma</u>: Given a test vector $u \in \mathbb{R}^n$ such that there is a vector $v \in \mathbb{R}^n$ with positive components and

$$M_\lambda^T u = - C^T v. \tag{6}$$

Then there exists no x such that

$$M_\lambda x = \Theta, \quad Cx \geqq \Theta \quad \text{and} \quad Cx \neq \Theta. \tag{7}$$

<u>Proof</u>: If there were such an x then

$$0 = x^T M_\lambda^T u = - x^T C^T v > 0,$$

a contradiction.

<u>Remark</u>. The assertion of the Lemma is the trivial part of Tucker's Theorem of the Alternative (see [4]).

By means of the Lemma the following Inclusion Theorem can be proved:

<u>Theorem 1</u>: Let C be nonsingular. For a test vector $u \in \mathbb{R}^n$ let

$$\Lambda(u) = \{\lambda \in \mathbb{R} \mid \text{Not all components of } C^{T-1} M_\lambda^T u \text{ have the same sign}\}. \tag{8}$$

Then the following implication is true:

$$M_\lambda x = \Theta,\ Cx \geqq \Theta,\ Cx \neq \Theta \quad \Longrightarrow \quad \lambda \in \Lambda(u). \tag{9}$$

Depending on C and M_λ it is more or less difficult to express the set $\Lambda(u)$ in explicit terms. For the linear eigenvalue problem it is possible to generate a large number of different inclusion theorems by choosing different matrix functions M_λ. In principle one can also find bounds for eigenvalues of nonlinear eigenvalue problems but this depends on the nonlinearity of the specific eigenvalue under consideration.

For the general eigenvalue problem [2]: Given (n,n)-matrices A and B, find λ and x such that

$$Ax = \lambda \cdot Bx, \tag{10}$$

one gets

Theorem 2: Let C be nonsingular and $u \in \mathbb{R}^n$ a test vector. Define

$$v = C^{T-1}A^T u, \quad w = C^{T-1}B^T u \tag{11}$$

and assume that w has only positive components ($w > \Theta$).
Let

$$\mu_i = v_i/w_i, \quad i = 1, \ldots, n. \tag{12}$$

If there is an eigenvalue λ of the general eigenvalue problem (10) with eigenvector x satisfying

$$Cx \geqq \Theta, \quad Cx \neq \Theta, \tag{13}$$

then

$$m := \min \mu_i \leqq \lambda \leqq \max \mu_i =: M. \tag{14}$$

For proving this relative of Collatz's Theorem, put $M_\lambda = A - \lambda \cdot B$. Note that no assumptions were made about A and B. Such assumptions, as in Collatz's Theorem, are necessary for guaranteeing existence of the eigenvalue. In order to have $w > \Theta$ one has to solve the inequality system $C^{T-1}B^T u > \Theta$ which is computationally hard except for simple cases (C = identity matrix and B = identity matrix or B nonnegative).

For $M_\lambda = (A - \lambda \cdot B)^T(A - \lambda \cdot B)$ one gets more interesting results:

Theorem 3: Let C be nonsingular and $u \in \mathbb{R}^n$ a test vector.
Define vectors $v, w \in \mathbb{R}^n$ by

$$v = \tfrac{1}{2} \cdot C^{T-1} \cdot (A^T B + B^T A)u, \tag{15}$$

$$w = C^{T-1}A^T Au \tag{16}$$

and assume that

$$z = C^{T-1}B^T Bu > \Theta. \tag{17}$$

Let

$$\mu_i = v_i/z_i, \quad \gamma_i = \mu_i^2 - w_i/z_i \tag{18}$$

and define intervals

$$\Lambda_i = \begin{cases} \emptyset & \text{for } \gamma_i < 0, \\ [\mu_i - \sqrt{\gamma_i},\ \mu_i + \sqrt{\gamma_i}] & \text{elsewhere,} \end{cases} \tag{19}$$

If there exists an eigenvalue λ of the general eigenvalue problem (10) with eigenvector x satisfying (13) then

$$\lambda \in \Lambda(u) = \bigcup \Lambda_i. \tag{20}$$

The numerical effort in applying the inclusion of Theorem 3 in the case of symmetric A and B = C = I (= identity matrix) as compared with Collatz's inclusion amounts to one additional multiplication of a matrix and a vector. For general A one has one more multiplication. However, since it is then possible to apply Collatz's inclusion to A as well to A^T, this additional effort pays by a better inclusion.

The set $\Lambda(u)$ of Theorem 3 for B = C = I and for symmetric A is generally very different from Collatz's interval (3). The numbers μ_i (18) are then identical with Collatz's numbers (2) but if $\Lambda(u)$ is not connected or if Λ_i belonging to $\mu_i = m$ or $\mu_i = M$ is empty, the set $\Lambda(u)$ has a completely different structure than the interval $[m, M]$. In the special case of an eigenvalue with nonnegative eigenvector the latter interval has an interesting property:

<u>Theorem 4</u>: Let B = C = I, $u \geqq \Theta$. For the Rayleigh quotient $R[u] = u^T Au/u^T u$ the inequality holds:

$$m < R[u] < M. \tag{21}$$

<u>Proof</u>: $R[u]$ is the minimizer of the quadratic functional $\phi(\lambda) = \|(A - \lambda \cdot I)u\|^2$. Since $u \geqq \Theta$, one has $\phi'(m) < 0$ and $\phi'(M) > 0$, hence $R[u] \in (m, M)$.

<u>Example 1</u>: Consider the matrix (Collatz, [1])

$$A = \begin{pmatrix} a + 2 & -1 \\ 1 & a - 2 \end{pmatrix}$$

with eigenvalues $\lambda_{1,2} = a \pm \sqrt{3}$ and eigenvectors $x_{1,2} = \begin{pmatrix} 1 \\ 2 \mp \sqrt{3} \end{pmatrix}$.
For $u = (1,1)^T$ we get $m = a - 1$ and $M = a + 1$
and none of the eigenvalues lies between m and M.
For matrix A^T we get $m^T = a - 3$, $M^T = a + 3$ and both eigenvalues

belong to this interval. Furthermore

$$\Lambda(u) = [a + 2 - \sqrt{3}, a + 2 + \sqrt{3}] \cup$$
$$\cup [a - 2 - \sqrt{3}, a - 2 + \sqrt{3}]$$

and $\lambda_{1,2} \in \Lambda(u)$. The Rayleigh quotient is $R[u] = a$ and $a \in [m, M]$, $a \in [m^T, M^T]$, however, $a \notin \Lambda(u)$.
For interpreting these results in the light of Theorem 3 note that A^T has no nonnegative eigenvectors:

$$x_{1,2}^{(T)} = (1, -2 \pm \sqrt{3})^T.$$

When an inclusion of an eigenvalue is found by means of Theorem 3 then it is possible to sharpen the bounds obtained:

- Take $\lambda_o \in \Lambda(u)$ and find some vector u with $C^{T-1}M_\lambda^T u > \Theta$. This approach amounts in solving a system of linear equations.
- For $C = B = I$ take $\lambda_o \in \Lambda(u)$ and try to reduce $\|M_{\lambda_o} u\|^2$ e.g. by means of coordinate relaxation (see [3]).

<u>Example 2</u>: Let

$$A = \begin{pmatrix} 1 & 2 & 0 & 0 \\ 0 & 1 & 2 & 0 \\ 0 & 0 & 1 & 2 \\ 1 & 0 & 0 & 1 \end{pmatrix}, \qquad B = C = I,$$

with eigenvalue $\lambda^* = 1 + \sqrt[4]{8} = 2.681793$ having nonnegative eigenvectors for A and A^T. The test vector $u = (1, 1, 1, 1)^T$ yields the inclusion according to Theorem 3

$$1.381 \leqq \lambda^* \leqq 3.619.$$

Now we put $\lambda = R[u] = 2.75$ and solve the linear system $M_\lambda v = (1,1,1,1)^T$. The solution is

$$v = \begin{pmatrix} 182.524 \\ 217.915 \\ 256.913 \\ 300.212 \end{pmatrix}$$

With test vector v one gets from Theorem 3:

$$\lambda^* \in [1.572, 2.748| \cup |2.757, 3.277].$$

Since in the proof of the Lemma only the trivial part of the Theorem of the Alternative was used, the whole procedure can be applied to fairly general eigenvalue problems. In general, however, it can be difficult to identify $\Lambda(u)$. This latter problem might be solved in a straitghtforward manner by means of an interval analytic approach (see e.g. [5]).

Example 3: Define the linear integral operator

$$\mathbb{K}u = \int_0^1 (s + t)\cdot u(s)\, ds.$$

We are looking for bounds for an eigenvalue belonging to a non-negative eigenfunction:

$$\mathbb{K}u = \lambda\cdot u, \qquad u(t) \geqq 0 \text{ for } 0 \leqq t \leqq 1.$$

with $u \equiv 1$ one has

$$(\mathbb{K}u - \lambda\cdot u)(t) = \frac{1}{2} + t - \lambda,$$

$$(M_\lambda u)(t) = ((\mathbb{K} - \lambda\cdot I)^*(\mathbb{K} - \lambda\cdot I)(u))(t) = $$

$$= \frac{7}{12} - \lambda + t - 2\lambda\cdot t + \lambda^2.$$

Then λ has to be determined such that

$$(M_\lambda u)(t) > 0 \quad \text{for all } t \in [0, 1].$$

In terms of interval analysis this reads [5]

$$\lambda^2 - \lambda\cdot[1,3] + [\tfrac{7}{12}, \tfrac{19}{12}] > 0$$

with solution

$$\lambda = \complement[-\ 0.790,\ 2.791].$$

By more careful estimation one gets

$$\Lambda(u) = [0.683,\ 2.317].$$

References

1. Collatz L (1942) Einschließungssatz für charakteristische Zahlen von Matrizen. Math. Z. 48:221-226
2. Collatz L (1963) Eigenwertaufgaben mit technischen Anwendungen. 2. Auflage. Akademische Verlagsgesellschaft Geest & Portig K.-G., Leipzig
3. Eckhardt U (1972) Iterative Lösung linearer Gleichungssysteme. Kernforschungsanlage Jülich, Jül-880-MA
4. Mangasarian OL (1969) Nonlinear Programming. McGraw-Hill Book Company, New York
5. Moore RE (1966) Interval Analysis. Prentice-Hall, Inc., Englewood Cliffs

Ulrich Eckhardt

Institut für Angewandte Mathematik

Bundesstraße 55

D-2000 Hamburg 13

Germany

ISNM, Vol. 69
Numerical Treatment of
Eigenvalue Problems, Vol. 3

EINE EINHEITLICHE HERLEITUNG VON EINSCHLIESSUNGSSÄTZEN FÜR EIGENWERTE

Friedrich Goerisch und Julius Albrecht

Symmetrisable eigenvalue problems of the form $M\phi = \lambda N\phi$ are considered. A simple method for obtaining inclusion theorems for eigenvalues is presented. It consists in combining a fundamental inclusion theorem with procedures for calculating upper bounds for the norm of the solutions of certain linear equations, $Nw = h$ or $M\tilde{w} = \tilde{h}$. Thus, numerous inclusion theorems are derived in a uniform manner.

Inhalt

Einleitung: Die Methode zur Herleitung von Einschließungssätzen

Für eine Klasse allgemeiner Eigenwertaufgaben $M\phi = \lambda N\phi$ wird eine einfache Methode zur Aufstellung von Einschließungssätzen für Eigenwerte geschildert. Sie besteht darin, Verfahren, welche die Berechnung oberer Schranken für die Norm der Lösung gewisser linearer Gleichungen $Nw = h$ oder $M\tilde{w} = \tilde{h}$ ermöglichen, mit einem grundlegenden Einschließungssatz zu kombinieren.

§1 Zunächst werden einige Voraussetzungen zusammengestellt, die für die gesamte Arbeit gelten.

Voraussetzung V1

1. $(H, \langle\cdot|\cdot\rangle)$ sei ein reeller Hilbertraum.
2. $D(M)$, $D(N)$ und $D(S)$ seien reelle Vektorräume mit $D(M) \subset D(N) \subset D(S)$.
3. $M: D(M) \to H$, $N: D(N) \to H$ und $S: D(S) \to H$ seien lineare Abbildungen.
4. Für alle $f,g \in D(N)$ gelte $\langle Sf|Ng\rangle = \langle Sg|Nf\rangle$; für alle $f \in D(N)$ mit $f \neq 0$ gelte $\langle Sf|Nf\rangle > 0$.
5. Für alle $f,g \in D(N^{-1}M)$ [1)] gelte $\langle Sf|Mg\rangle = \langle Sg|Mf\rangle$.

Ähnliche Voraussetzungen wurden von Schäfke und Schneider [48], [49] bei theoretischen Untersuchungen über Eigenwertaufgaben mit Systemen gewöhnlicher Differentialgleichungen verwendet. - Bei vielen Aufgaben ist $D(N) \subset H$, und dann kann meistens S als die identische Abbildung I gewählt werden. Hier wird $S = I$ jedoch nicht vorausgesetzt, so daß auch Eigenwertaufgaben mit Differentialgleichungen, bei denen der Eigenwert in den Randbedingungen auftritt, erfaßt werden können; dies soll an Hand einer Stekloffschen Eigenwertaufgabe erläutert werden:

Beispiel 1 [2)]

Eigenwertaufgabe: $\Delta\phi = 0$ in Ω, $\frac{\partial\phi}{\partial n} = \lambda\phi$ auf $\partial\Omega$

Einordnung: $H := L_2(\partial\Omega)$, $\langle f|g\rangle := \int_{\partial\Omega} fg\,ds$ für $f,g \in H$;

$$D(M) := D(N) := D(S) := \{f \in C^\infty(\overline{\Omega}) : \Delta f = 0 \text{ in } \Omega\},$$

$$Mf := \frac{\partial f}{\partial n}, \quad Nf := Sf := f_{|\partial\Omega} .$$

---

1) Ist P eine lineare Abbildung, so bezeichnet D(P) ihren Definitionsbereich.

2) Im folgenden bezeichnet Ω ein beschränktes, glattes (vgl. [5], S. 190, Def. 3-1) Gebiet im $\mathbb{R}^2$, $\partial\Omega$ den Rand von Ω und $\frac{\partial}{\partial n}$ die Ableitung in Richtung der äußeren Normalen von $\partial\Omega$.

Definition

Auf D(N) werden durch

$\{f|g\} := \langle Sf|Ng\rangle$ für $f,g \in D(N)$, $\|f\| := \sqrt{\langle Sf|Nf\rangle}$ für $f \in D(N)$

ein Skalarprodukt und eine Norm definiert.

Voraussetzung V2

Es gebe eine Folge $(\lambda_i)_{i\in\mathbb{N}}$ von Eigenwerten und eine Folge $(\phi_i)_{i\in\mathbb{N}}$ von Eigenelementen der Aufgabe $M\phi = \lambda N\phi$ mit der Eigenschaft, daß

$M\phi_i = \lambda_i N\phi_i$, $\{\phi_i|\phi_k\} = \delta_{ik}$ für $i,k \in \mathbb{N}$ (δ_{ik} Kronecker-Symbol)

und $\{f|f\} = \sum_{i=1}^{\infty} \{f|\phi_i\}^2$ für alle $f \in D(N^{-1}M)$ gilt.

Die Grundlage für alle weiteren Überlegungen bildet der folgende Einschließungssatz:

Satz 1

Es seien $\varrho,\tau \in \mathbb{R}$; U sei ein r-dimensionaler Unterraum von $D(N^{-1}M)$, und für alle $u \in U$ gelte $\|N^{-1}Mu - \varrho u\| \leq \tau\|u\|$.

Dann liegen mindestens r Eigenwerte[1)] der Aufgabe $M\phi = \lambda N\phi$ im Intervall $[\varrho-\tau,\varrho+\tau]$.

Beweis: Annahme: Das Intervall $[\varrho-\tau,\varrho+\tau]$ enthält höchstens r-1 Eigenwerte von $M\phi = \lambda N\phi$. E sei der Raum, der von denjenigen Eigenelementen aufgespannt wird, die zu den in $[\varrho-\tau,\varrho+\tau]$ gelegenen Eigenwerten gehören; da seine Dimension höchstens r-1 ist, gibt es Elemente $u \in U$ mit

$u \neq 0$, $\{u|\phi_i\} = 0$ für alle $i \in \mathbb{N}$ mit $\lambda_i \in [\varrho-\tau,\varrho+\tau]$.

Ein solches u sei fest gewählt; nach der Besselschen Ungleichung gilt

$$\sum_{i=1}^{\infty} (\lambda_i-\varrho)^2\{u|\phi_i\}^2 \leq \{N^{-1}Mu-\varrho u|N^{-1}Mu-\varrho u\},$$

und aus $\|N^{-1}Mu-\varrho u\| \leq \tau\|u\|$ folgt mit Hilfe der Voraussetzung V2

$$\sum_{\substack{i\in\mathbb{N} \\ \lambda_i \notin [\varrho-\tau,\varrho+\tau]}} (\lambda_i-\varrho)^2\{u|\phi_i\}^2 \leq \tau^2 \sum_{\substack{i\in\mathbb{N} \\ \lambda_i \notin [\varrho-\tau,\varrho+\tau]}} \{u|\phi_i\}^2 .$$

---

1) Eigenwerte werden stets entsprechend ihrer Vielfachheit gezählt.

Da $(\lambda_i-\varrho)^2 > \tau^2$ für alle $i \in \mathbb{N}$ mit $\lambda_i \notin [\varrho-\tau,\varrho+\tau]$ gilt, ergibt sich $\{u|\phi_i\}^2 = 0$ für alle $i \in \mathbb{N}$. Dies steht im Widerspruch zu $u \neq 0$.

Für bestimmte Zwecke ist eine andere Fassung des grundlegenden Einschließungssatzes etwas günstiger:

Satz 2

M sei injektiv. Es seien $\varrho,\tau \in \mathbb{R}$ mit $0 \leq \tau < |\varrho|$; V sei ein r-dimensionaler Unterraum von $D(M^{-1}N)$, und für alle $v \in V$ gelte

$$\|v - \frac{\varrho^2-\tau^2}{\varrho} M^{-1}Nv\| \leq \frac{\tau}{|\varrho|} \|v\|. \tag{1}$$

Dann liegen mindestens r Eigenwerte der Aufgabe $M\phi = \lambda N\phi$ im Intervall $[\varrho-\tau,\varrho+\tau]$.

Beweis (durch Zurückführung auf Satz 1): $U := \{M^{-1}Nv\colon v \in V\}$ ist ein r-dimensionaler Unterraum von $D(N^{-1}M)$. Gilt $u \in U$ und setzt man $v := N^{-1}Mu$, so folgt wegen $v \in V$ zunächst

$\{v - \frac{\varrho^2-\tau^2}{\varrho} u \mid v - \frac{\varrho^2-\tau^2}{\varrho} u\} \leq \frac{\tau^2}{\varrho^2}\{v|v\}$ und hieraus dann

$\{v-\varrho u|v-\varrho u\} \leq \tau^2\{u|u\}$, also $\|N^{-1}Mu-\varrho u\| \leq \tau \|u\|$.

Bemerkung: Sind die Voraussetzungen von Satz 2 erfüllt und ist ϱ positiv, so ergibt sich bei Verwendung der Abkürzungen

$$\lambda_* := \frac{\varrho^2-\tau^2}{\varrho}, \quad \varepsilon_* := \frac{\tau}{\varrho}$$

für das Einschließungsintervall die Darstellung

$$[\varrho-\tau,\varrho+\tau] = [\frac{\lambda_*}{1+\varepsilon_*}, \frac{\lambda_*}{1-\varepsilon_*}] .$$

§2 Die Einschließungssätze für Eigenwerte können folgendermaßen eingeteilt werden:

I Sätze, bei denen Paare (u,v) mit $u \in D(M)$, $v \in D(N)$, $Mu = Nv$ zur Einschließung benötigt werden,

II Sätze, bei deren Anwendung man zwar Elemente u aus $D(N^{-1}M)$ benutzen muß, $v := N^{-1}Mu$ aber nicht auszurechnen braucht, und

III Sätze, welche eine Bestimmung von Einschließungsintervallen ohne Verwendung von Elementen aus $D(N^{-1}M)$ ermöglichen.

Die Methode zur Herleitung von Einschließungssätzen besteht darin, eine der beiden Fassungen des grundlegenden Satzes aus §1 mit Abschätzungen für die Norm der Lösung linearer Gleichungen zu kombinieren. (Einen Überblick vermittelt ein Diagramm am Schluß der Arbeit.)

Soll ein unter II fallender Satz hergeleitet werden, so verwendet man Satz 1 zum Beweis, faßt die dort auftretende Größe $N^{-1}Mu - \varrho u$ als Lösung w der linearen Gleichung
$Nw = h$ (mit $h := Mu - \varrho Nu$)
auf und zieht zum Nachweis, daß die Voraussetzungen von Satz 1 erfüllt sind, ein Verfahren heran, das obere Schranken für $\|w\|$ liefert.
Wenn ein unter III fallender Satz hergeleitet werden soll, benutzt man Satz 2 zum Beweis, sieht die in (1) auftretende Größe $v - (\varrho^2-\tau^2)\varrho^{-1}M^{-1}Nv$ als Lösung $\tilde{w}$ der linearen Gleichung
$M\tilde{w} = \tilde{h}$ (mit $\tilde{h} := Mv - (\varrho^2-\tau^2)\varrho^{-1}Nv$)
an - dies ist natürlich nur unter der Zusatzvoraussetzung $v \in D(M)$ möglich - und macht beim Nachweis, daß (1) erfüllt ist, von Abschätzungen für $\|\tilde{w}\|$ Gebrauch.
Die unter I zusammengefaßten Sätze, die klassischen Einschließungssätze, werden nicht gesondert bewiesen, da sie sich unmittelbar als Spezialfälle der zu II gehörenden Sätze ergeben.

In der geschilderten Weise können zwar viele, aber nicht sämtliche Einschließungssätze hergeleitet werden. Resultate, die man nach der Methode der Zwischenaufgaben ("intermediate problems" [56]) oder nach der Methode der Orthogonalinvarianten [16] erhält, ordnen sich hier nicht ein.

Die Möglichkeit, Einschließungssätze in einheitlicher Weise herzuleiten, wurde zuerst von Wielandt [57] erkannt; er wies nach, daß der Satz von Kryloff-Bogoliubov, der Satz von Temple und Collatz sowie der Collatzsche Quotienten-Einschließungssatz für Matrizen aus einem grundlegenden Satz folgen. Dieser Gedanke wurde in verschiedenen Richtungen weiterentwickelt (z. B. [7], [15], [46]). In der vorliegenden Arbeit wird aufgezeigt, daß man sowohl die klassischen als auch die unter II und III einzuordnenden Einschließungssätze in einheitlicher Weise gewinnen kann, wenn man Normabschätzungen für die Lösung linearer Gleichungen als weiteren wesentlichen Bestandteil der Beweismethode hinzunimmt.

An Hand von zwei Beispielen soll nun erläutert werden, daß es nicht immer möglich und sinnvoll ist, Paare (u,v) mit der Eigenschaft

$$u \in D(M),\ v \in D(N),\ Mu = Nv \tag{2}$$

zur Eigenwerteinschließung zu verwenden, und daß es deshalb wichtig ist, auch andere als die klassischen Einschließungssätze zur Verfügung zu haben.

Beispiel 2

Eigenwertaufgabe: $\Delta^2\phi = \lambda(-\Delta\phi + 2d\,\frac{\partial^2\phi}{\partial x\partial y})$ in Ω (mit $d \in \mathbb{R}$, $|d| < 1$),

$$\phi = 0,\ \frac{\partial\phi}{\partial n} = 0 \qquad \text{auf } \partial\Omega$$

Einordnung: $H := L_2(\Omega)$, $\langle f|g\rangle := \int_\Omega fg\,dxdy$ für $f,g \in H$,

$D(M) := \{f \in C^\infty(\overline{\Omega}) : f = 0,\ \frac{\partial f}{\partial n} = 0 \text{ auf } \partial\Omega\}$,

$D(N) := \{f \in C^\infty(\overline{\Omega}) : f = 0 \text{ auf } \partial\Omega\}$, $\quad D(S) := H$,

$Mf := \Delta^2 f$, $\quad Nf := -\Delta f + 2d\,\frac{\partial^2 f}{\partial x\partial y}$, $\quad Sf := f$.

Die Bedingung (2) lautet hier

$$\Delta^2 u = -\Delta v + 2d\,\frac{\partial^2 v}{\partial x\partial y} \text{ in } \Omega,\quad u = 0,\ \frac{\partial u}{\partial n} = 0,\ v = 0 \text{ auf } \partial\Omega.$$

Wählt man ein $u \in D(M)$, so muß man v als exakte Lösung einer Randwertaufgabe bestimmen; Entsprechendes gilt, wenn man mit der Wahl eines $v \in D(N)$ beginnt.

Beispiel 3

Eigenwertaufgabe: $-\Delta\phi = \lambda\phi$ in Ω, $\phi = 0$ auf $\partial\Omega$

Einordnung: $H := L_2(\Omega) \times L_2(\partial\Omega)$; $\quad$ mit $\omega \in \mathbb{R}$, $\omega > 0$ sei

$$\left\langle \begin{pmatrix} f_1 \\ f_2 \end{pmatrix} \middle| \begin{pmatrix} g_1 \\ g_2 \end{pmatrix} \right\rangle := \int_\Omega f_1 g_1\,dxdy + \omega\int_{\partial\Omega} f_2 g_2\,ds \text{ für } \begin{pmatrix} f_1 \\ f_2 \end{pmatrix}, \begin{pmatrix} g_1 \\ g_2 \end{pmatrix} \in H,$$

$D(M) := D(N) := D(S) = C^\infty(\overline{\Omega})$,

$$Mf := \begin{pmatrix} -\Delta f \\ f_{|\partial\Omega} \end{pmatrix},\quad Nf := Sf := \begin{pmatrix} f \\ 0 \end{pmatrix}.$$

Paare (u,v), die der Bedingung (2) genügen, für die also

$-\Delta u = v$ in Ω, $u = 0$ auf $\partial\Omega$

gilt, wird man im allgemeinen leicht finden, schwierig kann es jedoch sein, die Paare (u,v) außerdem noch so zu wählen, daß sie sowohl eine für die Rechnung hinreichend einfache Gestalt haben als auch genaue Einschließungen liefern.

I Die klassischen Einschließungssätze für Aufgaben $M\phi = \lambda N\phi$

Zunächst werden Sätze zusammengestellt, bei denen Paare (u,v) mit der Eigenschaft $u \in D(M)$, $v \in D(N)$, $Mu = Nv$ zur Eigenwerteinschließung verwendet werden. Diese Sätze gehen - im Gegensatz zu den weiter unten besprochenen - im wesentlichen auf Veröffentlichungen aus der Zeit vor 1950 zurück und werden deshalb als klassische Einschließungssätze bezeichnet; sie sind Spezialfälle der in §§5 und 6 bewiesenen Sätze.

§3 Eine von Lehmann [37], [38], [39] entdeckte Verallgemeinerung des Satzes von Kryloff-Bogoliubov ermöglicht die Bestimmung von Einschließungsintervallen mit vorgeschriebenem Mittelpunkt ϱ:

Satz 3

Voraussetzungen und Bezeichnungen:

1. $u_1,\ldots,u_n$ seien linear unabhängige Elemente von $D(N^{-1}M)$; für $i=1,\ldots,n$ sei $v_i := N^{-1}Mu_i$.
2. Es seien $\varrho \in \mathbb{R}$,
   $A := (\{v_i|v_k\} - 2\varrho\{v_i|u_k\} + \varrho^2\{u_i|u_k\})_{i,k=1,\ldots,n}$ und
   $B := (\{u_i|u_k\})_{i,k=1,\ldots,n}$.
3. Für $i=1,\ldots,n$ sei μ_i der i-tkleinste Eigenwert der Matrix-Eigenwertaufgabe $Ax = \mu Bx$.

Behauptung: Für $r=1,\ldots,n$ liegen im Intervall $[\varrho-\sqrt{\mu_r},\varrho+\sqrt{\mu_r}]$ mindestens r Eigenwerte der Eigenwertaufgabe $M\phi = \lambda N\phi$.

Beweis: Die Behauptung ergibt sich aus Satz 14 mit Hilfe von (10).

Zur Berechnung von Einschließungsintervallen mit vorgeschriebenem rechten Endpunkt L dient die folgende ebenfalls auf Lehmann [37], [38], [39] zurückgehende Verallgemeinerung des Satzes von Temple und Collatz:

Satz 4

Voraussetzungen und Bezeichnungen:

1. $u_1,\ldots,u_n$ seien linear unabhängige Elemente von $D(N^{-1}M)$; für $i=1,\ldots,n$ sei $v_i := N^{-1}Mu_i$.
2. Es seien $L \in \mathbb{R}$,
   $A := (\{v_i|u_k\}-L\{u_i|u_k\})_{i,k=1,\ldots,n}$ und
   $B := (\{v_i|v_k\}-2L\{v_i|u_k\} + L^2\{u_i|u_k\})_{i,k=1,\ldots,n}$;
   B sei positiv definit.[1)]
3. Für $i=1,\ldots,n$ sei μ_i der i-tkleinste Eigenwert der Matrix-Eigenwertaufgabe $Ax = \mu Bx$.

Behauptung: Für alle $r \in \mathbb{N}$ mit $r \leq n$ und $\mu_r < 0$ enthält das Intervall $[L + \frac{1}{\mu_r}, L]$ mindestens r Eigenwerte der Eigenwertaufgabe $M\phi = \lambda N\phi$.

Beweis: Die Behauptung ergibt sich aus Satz 13 (für $t:=0$, $N_0:=N$).

Satz 4, der sich numerisch sehr gut bewährt hat (s. z. B. [3]), wird in [37], [38], [39] für den Fall D(S) = H, S = I bewiesen; für n = 1, D(S) = H, S = I reduziert er sich auf den Satz von Temple [54] und Collatz ([11], S. 174). Ergebnisse für den Fall, daß n = 1, aber S nicht notwendig die identische Abbildung ist, finden sich in zahlreichen Arbeiten ([4], [9], [13], [29]). Rektorys ([44], S. 493) hat den Satz von Temple und Collatz auf Eigenwertaufgaben in Bilinearformgestalt übertragen (vgl. auch [45]).

Zur Anwendung des nun folgenden Quotienten-Einschließungssatzes, der auf einem Grundgedanken von Collatz [10] beruht,

---

1) Ist L nicht Eigenwert der Aufgabe $M\phi = \lambda N\phi$, so ist B stets positiv definit.

braucht man - anders als bei den Sätzen 3 und 4 - keine Skalarprodukte auszurechnen; statt dessen muß man jedoch im allgemeinen Infima und Suprema reellwertiger Funktionen bestimmen.

Satz 5

Voraussetzungen und Bezeichnungen:

1. Auf $D(N)$ sei eine Halbordnung definiert, und zwar so, daß $(D(N), \leq)$ ein halbgeordneter linearer Raum ist und aus $f,g \in D(N)$, $f \geq 0$, $g \geq 0$ stets $\{f|g\} \geq 0$ folgt.
2. Es sei $u \in D(N^{-1}M)$ mit $u \neq 0$ und $u \geq 0$; es gebe $\underline{\alpha}, \overline{\alpha} \in \mathbb{R}$ mit $\underline{\alpha}u \leq N^{-1}Mu \leq \overline{\alpha}u$.
3. $\alpha_{min} := \sup\{\alpha \in \mathbb{R}: \alpha u \leq N^{-1}Mu\}$; $\alpha_{max} := \inf\{\alpha \in \mathbb{R}: N^{-1}Mu \leq \alpha u\}$.

Behauptung: Im Intervall $[\alpha_{min}, \alpha_{max}]$ liegt mindestens ein Eigenwert der Eigenwertaufgabe $M\phi = \lambda N\phi$.

Beweis: Die Behauptung ergibt sich aus Satz 10, wenn man dort $t := \frac{1}{2}(\alpha_{max}+\alpha_{min})$, $N_o := N$ setzt und berücksichtigt, daß

$\sup\{\beta \in \mathbb{R}: \beta u \leq N^{-1}Mu - tu\} = \alpha_{min} - t \leq 0$,

$\inf\{\beta \in \mathbb{R}: N^{-1}Mu - tu \leq \beta u\} = \alpha_{max} - t \geq 0$ gilt.

Satz 5 enthält als Spezialfall den in [1] für Eigenwertaufgaben mit gewöhnlichen Differentialgleichungen aufgestellten Einschließungssatz; der Kürze halber wird er hier nur für solche Aufgaben formuliert, bei denen alle Randbedingungen wesentlich sind:

Satz 6

Voraussetzungen und Bezeichnungen:

1. Es seien $m,n \in \mathbb{N}_o$ mit $m > n$; $x_o, x_1 \in \mathbb{R}$ mit $x_o < x_1$; $p_i \in C^i[x_o,x_1]$ für $i=0,\ldots,m$; $q_i \in C^i[x_o,x_1]$ für $i=0,\ldots,n$; $J_1, J_2 \subset \mathbb{N}_o$ mit $J_1 \cap J_2 = \emptyset$, $J_1 \cup J_2 = \{0,1,\ldots,n\}$, $n \in J_1$. Es gelte $p_m(x) > 0$ für $x \in [x_o,x_1]$; $q_i(x) > 0$ für $x \in [x_o,x_1]$, $i \in J_1$; $q_i(x) = 0$ für $x \in [x_o,x_1]$, $i \in J_2$.
2. Es seien
$D(M) := \{f \in C^{2m}[x_o,x_1]: f^{(k)}(x_o) = f^{(k)}(x_1) = 0 \text{ für } k=0,\ldots,m-1\}$,
$D(N) := \{f \in C^{2n}[x_o,x_1]: f^{(k)}(x_o) = f^{(k)}(x_1) = 0 \text{ für } k=0,\ldots,n-1\}$,
$Mf := \sum_{i=0}^{m} (-1)^i (p_i f^{(i)})^{(i)}$ und $Nf := \sum_{i=0}^{n} (-1)^i (q_i f^{(i)})^{(i)}$.

3. Es seien $u \in D(M)$ und $v \in D(N)$ mit $Mu = Nv$.
Für alle $x \in [x_o, x_1]$ mit Ausnahme von höchstens endlich vielen und für alle $i \in J_1$ gelte $u^{(i)}(x) \neq 0$. Es gebe $\underline{\alpha}, \overline{\alpha} \in \mathbb{R}$ mit der Eigenschaft, daß $\underline{\alpha} \leq \frac{v^{(i)}(x)}{u^{(i)}(x)} \leq \overline{\alpha}$ für alle $i \in J_1$ und alle $x \in [x_o, x_1]$ mit $u^{(i)}(x) \neq 0$ gilt.

4.
$$\alpha_{min} := \min_{i \in J_1} \inf_{\substack{x \in [x_o, x_1] \\ u^{(i)}(x) \neq 0}} \frac{v^{(i)}(x)}{u^{(i)}(x)} ; \qquad (3.1)$$

$$\alpha_{max} := \max_{i \in J_1} \inf_{\substack{x \in [x_o, x_1] \\ u^{(i)}(x) \neq 0}} \frac{v^{(i)}(x)}{u^{(i)}(x)} . \qquad (3.2)$$

Behauptung: Im Intervall $[\alpha_{min}, \alpha_{max}]$ liegt mindestens ein Eigenwert der Eigenwertaufgabe $M\phi = \lambda N\phi$.

Beweis: Es seien $H := L_2[x_o, x_1]$, $\langle f|g\rangle := \int_{x_o}^{x_1} fg\,dx$ für $f,g \in H$, $D(S) := D(N)$, $S := I$. Dann sind die Voraussetzungen V1 und V2 aus §1 erfüllt (bezüglich V2 s. [11], S. 141). Für $f,g \in D(N)$ sei $f \leq g$ gleichbedeutend damit, daß $f^{(i)}(x)u^{(i)}(x) \leq g^{(i)}(x)u^{(i)}(x)$ für alle $i \in J_1$ und alle $x \in [x_o, x_1]$ gilt. Aus Satz 5 ergibt sich nun die Behauptung.

Die Frage, für welches i das Minimum in (3.1) und das Maximum in (3.2) angenommen wird, ist von Held [27],[28] untersucht worden.
Im Spezialfall $J_1 = \{n\}$ - die Aufgabe $M\phi = \lambda N\phi$ gehört dann zur "Eingliedklasse" - reduziert sich Satz 6 auf einen von Collatz ([11], S. 132) bewiesenen Einschließungssatz.

Angemerkt sei noch, daß Quotienten-Einschließungssätze auch unter ganz anderen Voraussetzungen als den hier betrachteten bewiesen werden können; so wird beispielsweise in [12] unter anderem die Existenz einer nichtnegativen Eigenfunktion vorausgesetzt.

II Einschließung ohne Berechnung von $N^{-1}Mu$

Als nächstes werden Einschließungssätze, zu deren Anwendung man zwar Elemente u aus $D(N^{-1}M)$ benötigt, $N^{-1}Mu$ aber nicht auszurechnen braucht, aus dem grundlegenden Satz hergeleitet. Hierbei spielen Abschätzungen für die Norm der Lösung w einer linearen Gleichung

$$Nw = h \tag{4}$$

eine entscheidende Rolle. Die Norm von w kann man im allgemeinen dann leicht abschätzen, wenn

a) N ein Operator von monotoner Art ist (§4),

b) ein leicht invertierbarer Operator N_o bekannt ist, der in einem gewissen Sinn kleiner als N ist (§5), oder

c) die Lösung von (4) durch komplementäre Extremalprinzipien charakterisiert werden kann (§6).

§4 Wenn N (bezüglich geeigneter Halbordnungen in D(N) und H, s.u.) ein Operator von monotoner Art ist, erweist sich die folgende Abschätzung als nützlich:

Satz 7

Voraussetzungen und Bezeichnungen:

1. Auf H sei eine Halbordnung erklärt, und zwar so, daß $(H, \leq)$ ein halbgeordneter linearer Raum ist und aus $f, g \in H$, $f \geq 0$, $g \geq 0$ stets $\langle f|g\rangle \geq 0$ folgt.
2. Für alle $f, g \in D(N)$ mit $Nf \leq Ng$ gelte $Sf \leq Sg$.
3. Es seien $w, z \in D(N)$ und $h \in H$; es gelte $Nw = h$ und $-Nz \leq h \leq Nz$.

Behauptung: $\|w\| \leq \|z\|$

Beweis: Aus $-Nz \leq Nw \leq Nz$ folgt zunächst $-Sz \leq Sw \leq Sz$ und weiter $0 \leq \langle Nz + Nw \mid Sz - Sw\rangle = \{z|z\} - \{w|w\}$.

Bemerkung: Ist die Voraussetzung 1 von Satz 7 erfüllt, so kann man auf D(N) eine Halbordnung $\leq$ durch die Festsetzung definieren, daß $f \leq g$ gleichbedeutend mit $Sf \leq Sg$ ist. Die Voraussetzung 2 besagt dann gerade, daß $N : (D(N), \leq) \to (H, \leq)$ ein Operator von monotoner Art ist.

Kombiniert man die Abschätzung aus Satz 7 mit dem

grundlegenden Satz 1, so erhält man den folgenden Quotienten-Einschließungssatz:

Satz 8

Voraussetzungen und Bezeichnungen:

1. Die Voraussetzungen 1 und 2 von Satz 7 seien erfüllt.
2. Es sei $u \in D(N^{-1}M)$ mit $u \neq 0$ und $Nu \geq 0$; es gebe $\underline{\beta}, \overline{\beta} \in \mathbb{R}$ mit $\underline{\beta}Nu \leq Mu \leq \overline{\beta}Nu$.
3. $\beta_{min} := \sup\{\beta \in \mathbb{R} : \beta Nu \leq Mu\}$; $\beta_{max} := \inf\{\beta \in \mathbb{R} : Mu \leq \beta Nu\}$.

Behauptung: Im Intervall $[\beta_{min}, \beta_{max}]$ liegt mindestens ein Eigenwert der Aufgabe $M\phi = \lambda N\phi$.

Beweis: Da $Nu \geq 0$, $Nu \neq 0$ gilt, ist die Menge $\{\beta \in \mathbb{R}: \beta Nu \leq Mu\}$ durch $\overline{\beta}$ nach oben und die Menge $\{\beta \in \mathbb{R}: Mu \leq \beta Nu\}$ durch $\underline{\beta}$ nach unten beschränkt. Es sei $\varepsilon \in \mathbb{R}$ mit $\varepsilon > 0$; mit den Bezeichnungen $\varrho := \frac{1}{2}(\beta_{max} + \beta_{min})$ und $\tau := \frac{1}{2}(\beta_{max} - \beta_{min})$ ergibt sich

$$(\varrho-\tau-\varepsilon)Nu = (\beta_{min} - \varepsilon)Nu \leq Mu \leq (\beta_{max} + \varepsilon)Nu = (\varrho+\tau+\varepsilon)Nu,$$

also $-(\tau+\varepsilon)Nu \leq Mu - \varrho Nu \leq (\tau+\varepsilon)Nu$. Setzt man in Satz 7 $w := N^{-1}Mu - \varrho u$, $z := (\tau+\varepsilon)u$ und $h := Mu - \varrho Nu$, so erhält man $\|N^{-1}Mu - \varrho u\| \leq (\tau+\varepsilon)\|u\|$. Da dies für beliebiges positives ε gilt, ergibt sich $\|N^{-1}Mu - \varrho u\| \leq \tau\|u\|$. Die Behauptung folgt nun aus Satz 1.

Der Ausgangspunkt für die Untersuchungen [18],[19], die auf Satz 8 führten, war ein Problem, das Collatz ([11], S. 304) im Zusammenhang mit den von ihm entdeckten Quotienten-Einschließungssätzen aufgeworfen hat. - Einige mit Hilfe von Satz 8 gewonnene numerische Resultate für Eigenwertaufgaben mit gewöhnlichen Differentialgleichungen finden sich in [2], [47].

Ein auf Bartsch [6] zurückgehender Quotienten-Einschließungssatz (vgl. [26]) läßt sich in ähnlicher Weise wie Satz 8 aus dem grundlegenden Einschließungssatz herleiten.

§5 Die Norm der Lösung w einer linearen Gleichung $Nw = h$ kann man auch dann in einfacher Weise abschätzen, wenn man einen leicht invertierbaren Operator N_0 kennt, der im Sinne einer gewissen Halbordnung (s.u.) kleiner als N ist.

<u>Satz 9</u>

<u>Voraussetzungen:</u>

1. $D(N_o)$ sei ein reeller Vektorraum mit $D(N) \subset D(N_o) \subset D(S)$; $N_o: D(N_o) \to H$ sei eine lineare Abbildung.
2. Für alle $f,g \in D(N_o)$ gelte $\langle Sf|N_o g\rangle = \langle Sg|N_o f\rangle$; für alle $f \in D(N_o)$ mit $f \neq 0$ gelte $\langle Sf|N_o f\rangle > 0$.
3. Für alle $f \in D(N)$ gelte $\langle Sf|N_o f\rangle \leq \langle Sf|Nf\rangle$.
4. Auf $D(N_o)$ sei eine Halbordnung definiert, und zwar so, daß $(D(N_o), \leq)$ ein halbgeordneter linearer Raum ist und aus $f,g \in D(N_o)$, $f \geq 0$, $g \geq 0$ stets $\langle Sf|N_o g\rangle \geq 0$ folgt.
5. Es seien $\alpha \in \mathbb{R}$, $w,z \in D(N)$ und $h \in H$; es gelte $|\alpha| \leq 1$, $Nw = h$, $h + \alpha Nz \in D(N_o^{-1})$ und $-z \leq N_o^{-1}(h + \alpha Nz) - \alpha z \leq z$.

<u>Behauptung:</u> $\|w\| \leq \|z\|$

<u>Beweis:</u> Aus $(1-\alpha)z + N_o^{-1}(h+\alpha Nz) \geq 0$ und $(1+\alpha)z - N_o^{-1}(h+\alpha Nz) \geq 0$ folgt

$$\langle S((1-\alpha)z + N_o^{-1}(h+\alpha Nz)) \mid N_o((1+\alpha)z - N_o^{-1}(h+\alpha Nz))\rangle \geq 0 . \tag{5}$$

Da $\alpha^2 \leq 1$ gilt, erhält man

$$(1-\alpha^2)\langle Sz|(N-N_o)z\rangle \geq 0. \tag{6}$$

Für alle $\begin{pmatrix} f_1 \\ f_2 \end{pmatrix}, \begin{pmatrix} g_1 \\ g_2 \end{pmatrix} \in D(N_o) \times D(N)$ sei

$$b(\begin{pmatrix} f_1 \\ f_2 \end{pmatrix}, \begin{pmatrix} g_1 \\ g_2 \end{pmatrix}) := \langle Sf_1|N_o g_1\rangle + \langle Sf_2|(N-N_o)g_2\rangle;$$

ferner sei $\hat{z} := \begin{pmatrix} N_o^{-1}(h+\alpha Nz) - \alpha z \\ -\alpha z \end{pmatrix}$.

Mit Hilfe von (5) und (6) ergibt sich

$$0 \leq b(\begin{pmatrix} z \\ z \end{pmatrix} + \hat{z}, \begin{pmatrix} z \\ z \end{pmatrix} - \hat{z}) = \{z|z\} - b(\hat{z},\hat{z}).$$

Da $b(\cdot,\cdot)$ eine positiv semidefinite Bilinearform ist, gilt

$$0 \leq b(\hat{z} - \begin{pmatrix} w \\ w \end{pmatrix}, \hat{z} - \begin{pmatrix} w \\ w \end{pmatrix}) = b(\hat{z},\hat{z}) - 2\langle Sw|h\rangle + \langle Sw|Nw\rangle$$

$$= b(\hat{z},\hat{z}) - \{w|w\}.$$

Insgesamt erhält man also $0 \leq \{z|z\} - \{w|w\}$.

Bemerkung: Auf der Menge aller linearen Abbildungen $P: D(P) \to H$ mit der Eigenschaft, daß $D(P) \subset D(S)$ und $\langle Sf|Pg\rangle = \langle Sg|Pf\rangle$ für alle $f,g \in D(P)$ gilt, kann man eine Halbordnung $\leq$ durch die Festlegung

$$\left.\begin{array}{l} P_1 \leq P_2 \Leftrightarrow \\ D(P_2) \subset D(P_1),\ \langle Sf|P_1 f\rangle \leq \langle Sf|P_2 f\rangle \text{ für alle } f \in D(P_2) \end{array}\right\} \quad (7)$$

definieren. Ist speziell $D(S) = H$, $S = I$, so stimmt diese Halbordnung mit der für symmetrische Operatoren üblichen überein. Der Operator N_o ist also im Sinne der soeben erklärten Halbordnung kleiner als N, wenn die Voraussetzungen von Satz 9 erfüllt sind.

Kombiniert man die Abschätzung aus Satz 9 mit dem grundlegenden Satz 1, so erhält man

Satz 10

Voraussetzungen und Bezeichnungen:

1. Die Voraussetzungen 1 bis 4 von Satz 9 seien erfüllt.
2. Es seien $t \in \mathbb{R}$ und $u \in D(N^{-1}M)$; es gelte $u \neq 0$, $u \geq 0$ und $Mu - tNu \in D(N_o^{-1})$. Es gebe $\underline{\beta}, \overline{\beta} \in \mathbb{R}$ mit $\underline{\beta}u \leq N_o^{-1}(Mu-tNu) \leq \overline{\beta}u$.
3. $\beta_{min} := \sup\{\beta \in \mathbb{R}: \beta \leq 0,\ \beta u \leq N_o^{-1}(Mu-tNu)\}$;
   $\beta_{max} := \inf\{\beta \in \mathbb{R}: \beta \geq 0,\ N_o^{-1}(Mu-tNu) \leq \beta u\}$.

Behauptung: Im Intervall $[\beta_{min}+t,\ \beta_{max}+t]$ liegt mindestens ein Eigenwert der Eigenwertaufgabe $M\phi = \lambda N\phi$.

Beweis: Es sei $\varepsilon \in \mathbb{R}$ mit $\varepsilon > 0$; zur Abkürzung wird $\sigma := \frac{1}{2}(\beta_{max} - \beta_{min}) + \varepsilon$ gesetzt. Die in Voraussetzung 5 von Satz 9 auftretenden Größen werden folgendermaßen definiert:
$\alpha := \frac{1}{2\sigma}(\beta_{max} + \beta_{min})$, $w := N^{-1}Mu - (t+\alpha\sigma)u$, $z := \sigma u$,
$h := Mu - (t+\alpha\sigma)Nu$.
Dann gilt $\sigma(\alpha-1) = \beta_{min} - \varepsilon < 0$ und $\sigma(\alpha+1) = \beta_{max} + \varepsilon > 0$; hieraus erhält man $\alpha^2 - 1 < 0$. Da $u \geq 0$ ist, ergibt sich
$\sigma(\alpha-1)u = (\beta_{min} - \varepsilon)u \leq N_o^{-1}(Mu-tNu) \leq (\beta_{max} + \varepsilon)u = \sigma(\alpha+1)u$,
also $(\alpha-1)z \leq N_o^{-1}(h+\alpha Nz) \leq (\alpha+1)z$. Nach Satz 9 folgt
$\|N^{-1}Mu - (t+\frac{1}{2}(\beta_{max} + \beta_{min}))u\| \leq \frac{1}{2}(\beta_{max} - \beta_{min} + 2\varepsilon)\|u\|$.

Wenn man hierin ε gegen 0 gehen läßt, erhält man mit Hilfe von Satz 1 die Behauptung.

Der für die Anwendung wichtigste Spezialfall (s. [18]) $D(S) = H$, $S = I$, $N_o = \tilde{\kappa}I$ (mit $\tilde{\kappa} \in \mathbb{R}$) wurde an Hand von verschiedenen Eigenwertaufgaben mit gewöhnlichen Differentialgleichungen numerisch erprobt [47]. - Setzt man $N_o := N$, so reduziert sich Satz 10 auf den klassischen Quotienten-Einschließungssatz (Satz 5).
Die Kenntnis eines leicht invertierbaren Operators N_o, der im Sinne der Halbordnung (7) kleiner als N ist, ermöglicht auch die Aufstellung von komplementären Extremalprinzipien und kann auf diese Weise zur Herleitung eines weiteren Einschließungssatzes (Satz 13) verwendet werden.

§6 Die folgende Abschätzung für die Norm der Lösung w von (4) hängt eng mit komplementären Extremalprinzipien zusammen:

Satz 11

Voraussetzungen:

1. X sei ein reeller Vektorraum.
2. $b(\cdot,\cdot)$ sei eine symmetrische Bilinearform auf X; für alle $f \in X$ gelte $b(f,f) \geq 0$.
3. $T : D(N) \to X$ sei eine lineare Abbildung; für alle $f,g \in D(N)$ gelte $b(Tf,Tg) = \{f|g\}$.
4. Es seien $w \in D(N)$ und $h \in H$ mit $Nw = h$.
5. Es sei $\tilde{z} \in X$; für alle $g \in D(N)$ gelte $b(Tg,\tilde{z}) = \langle Sg|h\rangle$.

Behauptung: $\|w\|^2 \leq b(\tilde{z},\tilde{z})$

Beweis: $0 \leq b(\tilde{z}-Tw,\tilde{z}-Tw) = b(\tilde{z},\tilde{z}) - 2\langle Sw|h\rangle + \{w|w\}$
$= b(\tilde{z},\tilde{z}) - \{w|w\}$.

Bemerkung: Sind die Voraussetzungen 1 bis 4 von Satz 11 erfüllt, so läßt sich w durch komplementäre Extremalprinzipien charakterisieren:

$$\max_{f\in D(N)} (2\langle Sf|h\rangle - \{f|f\}) = \|w\|^2 = \min_{\tilde{z}\in\tilde{X}} b(\tilde{z},\tilde{z}) \qquad (8)$$

mit $\tilde{X} := \{\tilde{z} \in X\colon b(Tg,\tilde{z}) = \langle Sg|h\rangle \text{ für alle } g \in D(N)\}$;

das Maximum wird dabei für $f = w$ angenommen. Die linke Gleichung in (8) ergibt sich daraus, daß
$0 \leq \{w-f|w-f\} = \{w|w\} - 2\langle Sf|h\rangle + \{f|f\}$ für alle $f \in D(N)$
gilt, die rechte folgt aus Satz 11 wegen $Tw \in \tilde{X}$. - Viele bekannte (vgl. [55]) komplementäre Extremalprinzipien lassen sich in die Form (8) bringen und ermöglichen es, $\|w\|$ geeignet abzuschätzen.

Aus den Sätzen 1 und 11 erhält man

Satz 12

Voraussetzungen und Bezeichnungen:

1. Die Voraussetzungen 1 bis 3 von Satz 11 seien erfüllt.
2. Es seien $u_1,\ldots,u_n \in D(N^{-1}M)$ und $v_1,\ldots,v_n \in X$; $u_1,\ldots,u_n$ seien linear unabhängig; für alle $g \in D(N)$ und $i=1,\ldots,n$ gelte
   $b(Tg,v_i) = \langle Sg|Mu_i\rangle$.
3. Es seien $L \in \mathbb{R}$,
   $A := (\langle Su_i|Mu_k\rangle - L\langle Su_i|Nu_k\rangle)_{i,k=1,\ldots,n}$ und
   $B := (b(v_i,v_k) - 2L\langle Su_i|Mu_k\rangle + L^2\langle Su_i|Nu_k\rangle)_{i,k=1,\ldots,n}$;
   B sei positiv definit[1].
4. Für $i=1,\ldots,n$ sei μ_i der i-tkleinste Eigenwert der Matrix-Eigenwertaufgabe $Ax = \mu Bx$.

Behauptung: Für alle $r \in \mathbb{N}$ mit $r \leq n$ und $\mu_r < 0$ enthält das Intervall $[L + \frac{1}{\mu_r}, L]$ mindestens r Eigenwerte der Eigenwertaufgabe $M\phi = \lambda N\phi$.

Beweis: Es sei $r \in \mathbb{N}$ mit $r \leq n$ und $\mu_r < 0$. $\tilde{U}$ sei der Unterraum des $\mathbb{R}^n$, der von denjenigen Eigenvektoren der Aufgabe $Ax = \mu Bx$ aufgespannt wird, die zu einem Eigenwert $\mu_i \leq \mu_r$ gehören. Dann gilt $\dim \tilde{U} \geq r$ und $c'Ac \leq \mu_r c'Bc$ für alle $c \in \tilde{U}$.
$U := \{\sum_{i=1}^{n} c_i u_i : (c_1,\ldots,c_n)' \in \tilde{U}\}$ ist ein Unterraum von $D(N^{-1}M)$ mit $\dim U \geq r$.
Es sei nun $u \in U$. Mit den eindeutig bestimmten Koeffizienten c_i

---

[1] Ist L nicht Eigenwert der Aufgabe $M\phi = \lambda N\phi$, so ist B stets positiv definit.

der Darstellung $u = \sum_{i=1}^{n} c_i u_i$ wird $v := \sum_{i=1}^{n} c_i v_i$ definiert. Dann gilt

$$\langle Su|Mu\rangle - L\langle Su|Nu\rangle \leq \mu_r(b(v,v) - 2L\langle Su|Mu\rangle + L^2\langle Su|Nu\rangle).$$

Unter Berücksichtigung der Beziehungen $\langle Su|Nu\rangle = b(Tu,Tu)$ und $\langle Su|Mu\rangle = b(Tu,v)$ ergibt sich hieraus

$$b(v - (\frac{1}{2\mu_r}+L)Tu,\ v - (\frac{1}{2\mu_r}+L)Tu) \leq (\frac{1}{2\mu_r})^2\{u|u\}. \tag{9}$$

Setzt man in Satz 11 $w := N^{-1}Mu - (\frac{1}{2\mu_r}+L)u$, $h := Mu - (\frac{1}{2\mu_r}+L)Nu$ und $\tilde{z} := v - (\frac{1}{2\mu_r}+L)Tu$, so erhält man

$$\|N^{-1}Mu - (\frac{1}{2\mu_r}+L)u\|^2 \leq b(v - (\frac{1}{2\mu_r}+L)Tu,\ v - (\frac{1}{2\mu_r}+L)Tu).$$

Mit Hilfe von (9) folgt hieraus

$$\|N^{-1}Mu - (\frac{1}{2\mu_r}+L)u\| \leq |\frac{1}{2\mu_r}|\|u\|.$$

Satz 1 liefert nun unmittelbar die Behauptung.

Der Spezialfall

$$\left.\begin{array}{ll} X := D(N),\ Tf := f & \text{für } f \in D(N), \\ b(f,g) := \langle Sf|Ng\rangle & \text{für } f,g \in D(N) \end{array}\right\} \tag{10}$$

von Satz 12 stimmt mit dem Satz von Temple-Collatz-Lehmann (Satz 4) überein. Bei geeigneter Wahl von X, b und T ist Satz 12 jedoch leichter zu handhaben als Satz 4 und besitzt einen größeren Anwendungsbereich.

Zur Behandlung von Beispiel 2 aus §2 kann man

$X := C^{\infty}(\bar{\Omega})$, $Tf := f$ für $f \in D(N)$,

$$b(f,g) := \int_{\Omega} (\frac{\partial f}{\partial x}\frac{\partial g}{\partial x} + \frac{\partial f}{\partial y}\frac{\partial g}{\partial y} + 2d\frac{\partial f}{\partial x}\frac{\partial g}{\partial y})dxdy \text{ für } f,g \in X$$

wählen; die in Satz 12 auftretenden Funktionen v_i brauchen dann lediglich den Bedingungen

$$v_i \in C^{\infty}(\bar{\Omega}),\quad -\Delta v_i + 2d\frac{\partial^2 v_i}{\partial x \partial y} = \Delta^2 u_i \quad \text{in } \Omega$$

zu genügen, also keine Randbedingungen zu erfüllen.

Es empfiehlt sich, für die Funktionen v_i einen Ansatz, der noch freie Parameter enthält, zu machen; die Gleichungen für die optimale Wahl dieser Parameter stimmen dann mit den Trefftzschen Gleichungen überein (s. [19], [20]). Numerisch erprobt wurde Satz 12 an Hand von Beispiel 2 aus §2 [36]. - Vor kurzem [25] wurde ein zu Satz 12 analoger Satz, bei dem nicht N, sondern M als positiv definit vorausgesetzt wird, bewiesen; er hat sich bei zahlreichen Aufgaben, z. B. bei verschiedenen Stekloffschen Eigenwertaufgaben [23] und bei einem System partieller Differentialgleichungen [21], gut bewährt.

Satz 12 kann durch geeignete Wahl von X, b und T auch an die Situation, daß man einen leicht invertierbaren Operator N_o kennt, der im Sinne der Halbordnung (7) kleiner als N ist, angepaßt werden; man erhält dann

Satz 13

Voraussetzungen und Bezeichnungen:

1. Die Voraussetzungen 1 bis 3 von Satz 9 seien erfüllt.
2. Es seien $u_1,\dots,u_n \in D(N^{-1}M) \cap D(N_o^{-1}M) \cap D(N_o^{-1}N)$.
3. Es seien $L,t \in \mathbb{R}$,

   $A := (\langle Su_i|Mu_k\rangle - L\langle Su_i|Nu_k\rangle)_{i,k=1,\dots,n}$ und

   $B := (\langle SN_o^{-1}(Mu_i-tNu_i) \mid Mu_k-tNu_k\rangle - 2(L-t)\langle Su_i|Mu_k\rangle$
   $+ (L^2-t^2)\langle Su_i|Nu_k\rangle)_{i,k=1,\dots,n}$;

   B sei positiv definit.
4. Für i=1,...,n sei μ_i der i-tkleinste Eigenwert der Matrix-Eigenwertaufgabe $Ax = \mu Bx$.

Behauptung: Für alle $r \in \mathbb{N}$ mit $r \leq n$ und $\mu_r < 0$ enthält das Intervall $[L + \frac{1}{\mu_r}, L]$ mindestens r Eigenwerte der Eigenwertaufgabe $M\phi = \lambda N\phi$.

Beweis: Setzt man $X := D(N_o) \times D(N)$,

$$b\left(\begin{pmatrix} f_1 \\ f_2 \end{pmatrix}, \begin{pmatrix} g_1 \\ g_2 \end{pmatrix}\right) := \langle Sf_1|N_o g_1\rangle + \langle Sf_2|(N-N_o)g_2\rangle \text{ für alle}$$

$$\begin{pmatrix} f_1 \\ f_2 \end{pmatrix}, \begin{pmatrix} g_1 \\ g_2 \end{pmatrix} \in X, \quad Tf := \begin{pmatrix} f \\ f \end{pmatrix} \text{ für alle } f \in D(N) \text{ und}$$

$$v_i := \begin{pmatrix} N_o^{-1}(Mu_i-tNu_i) + tu_i \\ tu_i \end{pmatrix} \text{ für } i=1,\dots,n, \text{ so erhält man die}$$

Behauptung unmittelbar aus Satz 12.

Geeignete Operatoren N_o kann man sich beispielsweise dann verschaffen, wenn

α) der Spezialfall $D(S) = H$, $S = I$ vorliegt,

β) eine Orthonormalbasis $(\psi_i)_{i\in\mathbb{N}}$ des Hilbertraums $(H,\langle\cdot|\cdot\rangle)$ bekannt ist, die aus Eigenelementen der Eigenwertaufgabe $N\psi = \kappa\psi$ besteht, und

γ) die Folge $(\kappa_i)_{i\in\mathbb{N}}$ der zu den Eigenelementen ψ_i gehörenden Eigenwerte κ_i der Aufgabe $N\psi = \kappa\psi$ monoton nicht fallend ist.

Man kann dann jeden der Operatoren $N_{o,j}: H \to H$,

$$N_{o,j}f := \sum_{i=1}^{j} \kappa_i \langle f|\psi_i\rangle\psi_i + \kappa_{j+1}(f - \sum_{i=1}^{j} \langle f|\psi_i\rangle\psi_i) \quad (j\in\mathbb{N}) \tag{11}$$

als Abbildung N_o in Satz 13 verwenden; $N_{o,j}^{-1}$ läßt sich unmittelbar angeben:

$$N_{o,j}^{-1}f = \sum_{i=1}^{j} \frac{1}{\kappa_i}\langle f|\psi_i\rangle\psi_i + \frac{1}{\kappa_{j+1}}(f - \sum_{i=1}^{j} \langle f|\psi_i\rangle\psi_i) \quad (j\in\mathbb{N}).$$

Operatoren der Form (11) wurden von Bazley und Fox [8] im Rahmen der Methode der Zwischenaufgaben ("intermediate problems") eingeführt. Der Vorschlag, solche Operatoren im Zusammenhang mit Einschließungssätzen zu benutzen, stammt von Schellhaas [50], [51], auf den auch zwei Spezialfälle von Satz 13 zurückgehen. Ferner hat Schellhaas analoge Resultate für Eigenwertaufgaben $M\phi = \lambda N\phi$, bei denen nicht N, sondern M positiv definit ist, angegeben und zur Berechnung von Beulwerten von Platten verwendet ([50], [51]).

Während man in Satz 12 den rechten Endpunkt L der Einschließungsintervalle frei wählen kann, ermöglicht der nun folgende Satz die Bestimmung von Einschließungsintervallen mit vorgeschriebenem Mittelpunkt ϱ; ansonsten sind die beiden Sätze sehr ähnlich, da sie beide auf der Normabschätzung aus Satz 11 beruhen.

Satz 14

Voraussetzungen und Bezeichnungen:

1. Die Voraussetzungen 1 bis 3 von Satz 11 und die Voraussetzung 2 von Satz 12 seien erfüllt.

2. Es seien $\varrho \in \mathbb{R}$,

$A := (b(v_i,v_k) - 2\varrho\langle Su_i|Mu_k\rangle + \varrho^2\langle Su_i|Nu_k\rangle)_{i,k=1,\ldots,n}$ und

$B := (\{u_i|u_k\})_{i,k=1,\ldots,n}$.

3. Für $i=1,\ldots,n$ sei μ_i der i-tkleinste Eigenwert der Matrix-Eigenwertaufgabe $Ax = \mu Bx$.

Behauptung: Für $r=1,\ldots,n$ liegen im Intervall $[\varrho - \sqrt{\mu_r}, \varrho + \sqrt{\mu_r}]$ mindestens r Eigenwerte der Eigenwertaufgabe $M\phi = \lambda N\phi$.

Beweis: Es sei $r \in \mathbb{N}$ mit $r \leq n$. $\tilde{U}$ sei der Unterraum des $\mathbb{R}^n$, der von denjenigen Eigenvektoren der Aufgabe $Ax = \mu Bx$ aufgespannt wird, die zu einem Eigenwert $\mu_i \leq \mu_r$ gehören. Dann gilt $\dim \tilde{U} \geq r$ und $c'Ac \leq \mu_r c'Bc$ für alle $c \in \tilde{U}$.

$U := \{ \sum_{i=1}^{n} c_i u_i : (c_1,\ldots,c_n)' \in \tilde{U}\}$ ist ein Unterraum von $D(N^{-1}M)$ mit $\dim U \geq r$.

Es sei nun $u \in U$. Mit den eindeutig bestimmten Koeffizienten c_i der Darstellung $u = \sum_{i=1}^{n} c_i u_i$ wird $v := \sum_{i=1}^{n} c_i v_i$ definiert. Dann gilt

$b(v,v) - 2\varrho\langle Su|Mu\rangle + \varrho^2\langle Su|Nu\rangle \leq \mu_r\{u|u\}$.

Unter Berücksichtigung der Beziehungen $\langle Su|Nu\rangle = b(Tu,Tu)$ und $\langle Su|Mu\rangle = b(Tu,v)$ ergibt sich hieraus $b(v-\varrho Tu,v-\varrho Tu) \leq \mu_r\{u|u\}$. Setzt man in Satz 11 $w := N^{-1}Mu - \varrho u$, $h := Mu - \varrho Nu$ und $\tilde{z} := v - \varrho Tu$, so erhält man $\|N^{-1}Mu - \varrho u\|^2 \leq b(v-\varrho Tu,v-\varrho Tu)$; aus Satz 1 folgt dann die Behauptung.

Der Satz von Kryloff-Bogoliubov-Lehmann (Satz 3) ergibt sich unmittelbar aus Satz 14, wenn man X, b und T wie in (10) wählt. Die Sätze 12, 13 und 14 enthalten ferner einige Einschließungssätze, die eng mit den von Falk [15] für normale Matrizenpaare aufgestellten zusammenhängen, als Spezialfälle.

III Einschließung ohne Verwendung von Elementen aus $D(N^{-1}M)$

Im folgenden werden Einschließungssätze, zu deren Anwendung man keine Elemente von $D(N^{-1}M)$ benötigt, aus Satz 2, der zweiten Fassung des grundlegenden Einschließungssatzes, hergeleitet. Hierzu werden Abschätzungen für die Norm der Lösung $\tilde{w}$ der Gleichung

$$M\tilde{w} = \tilde{h} \tag{12}$$

herangezogen, und zwar

a) Abschätzungen, die auf der monotonen Art von M beruhen (§7), und

b) a-priori-Abschätzungen (§8).

§7 Wenn M von monotoner Art ist, kann man im allgemeinen leicht obere Schranken für die Norm der Lösung $\tilde{w}$ der Gleichung (12) erhalten.

Satz 15

Voraussetzungen:

1. Auf $D(M)$ sei eine Halbordnung erklärt, und zwar so, daß $(D(M), \leq)$ ein halbgeordneter linearer Raum ist und aus $f, g \in D(M)$, $f \geq 0$, $g \geq 0$ stets $\{f|g\} \geq 0$ folgt.
2. Auf H sei eine Halbordnung $\leq$ erklärt; M sei ein Operator monotoner Art, d. h. aus $f, g \in D(M)$ und $Mf \leq Mg$ folgt stets $f \leq g$.
3. Es seien $\tilde{w}, z \in D(M)$ und $\tilde{h} \in H$; es gelte $M\tilde{w} = \tilde{h}$ und $-Mz \leq \tilde{h} \leq Mz$.

Behauptung: $\|\tilde{w}\| \leq \|z\|$

Beweis: Aus $-Mz \leq \tilde{h} \leq Mz$ folgt zunächst $-z \leq \tilde{w} \leq z$ und weiter $0 \leq \{z-\tilde{w}|z+\tilde{w}\} = \{z|z\} - \{\tilde{w}|\tilde{w}\}$.

Kombiniert man diese Abschätzung mit Satz 2, so erhält man den folgenden Einschließungssatz:

Satz 16

Voraussetzungen und Bezeichnungen:

1. Die Voraussetzungen 1 und 2 von Satz 15 seien erfüllt.
2. Es seien $\varepsilon_*, \lambda_* \in \mathbb{R}$, $v \in D(M^{-1}N) \cap D(M)$ und $z \in D(M)$;

es gelte $0 \leq \varepsilon_* < 1$, $\lambda_* > 0$, $v \neq 0$, $-Mz \leq Mv - \lambda_* Nv \leq Mz$ und $\|z\| \leq \varepsilon_* \|v\|$.

Behauptung: Im Intervall $[\frac{\lambda_*}{1+\varepsilon_*}, \frac{\lambda_*}{1-\varepsilon_*}]$ liegt mindestens ein Eigenwert der Eigenwertaufgabe $M\phi = \lambda N\phi$.

Beweis: Setzt man in Satz 15 $\tilde{w} := v-\lambda_* M^{-1}Nv$ und $\tilde{h} := Mv-\lambda_* Nv$, so erhält man $\|v-\lambda_* M^{-1}Nv\| \leq \|z\| \leq \varepsilon_* \|v\|$; hieraus folgt nach Satz 2 die Behauptung.

Eine einfache Möglichkeit, Satz 16 auf Beispiel 3 aus §2 anzuwenden, sei kurz skizziert: D(M) und H werden mit der kanonischen Halbordnung versehen; $\lambda_* \in \mathbb{R}$ und $v \in C^\infty(\overline{\Omega})$ werden so gewählt, daß die Bedingungen $\lambda_* > 0$, $-\Delta v = \lambda_* v$ und $\int_\Omega dxdy \leq \int_\Omega v^2 dxdy$ erfüllt sind; ε_* und z werden durch $\varepsilon_* := \max_{(x,y)\in\partial\Omega} |v(x,y)|$ und $z(x,y) := \varepsilon_*$ für alle $(x,y) \in \overline{\Omega}$ definiert. Gilt nun $\varepsilon_* < 1$, so enthält das Intervall $[\frac{\lambda_*}{1+\varepsilon_*}, \frac{\lambda_*}{1-\varepsilon_*}]$ mindestens einen Eigenwert der betrachteten Aufgabe.

Die Idee, die Satz 16 zugrunde liegt, stammt von Fox, Henrici und Moler [17]; sie wurde von Nickel [43], Moler und Payne [42] sowie Donnelly [14] weiterentwickelt. Numerische Resultate liegen zu Membranproblemen für verschiedene Grundgebiete vor ([14], [17]).

§8 Kennt man eine positive Zahl γ derart, daß

$$\{f|f\} \leq \gamma\langle Mf|Mf\rangle \quad \text{für alle } f \in D(M) \tag{13}$$

gilt, so kann man leicht eine obere Schranke für die Norm der Lösung $\tilde{w}$ der Gleichung (12) angeben:

$$\|\tilde{w}\|^2 \leq \gamma\langle\tilde{h}|\tilde{h}\rangle . \tag{14}$$

Wenn man diese Abschätzung mit Satz 2 kombiniert, erhält man

Satz 17

Voraussetzungen und Bezeichnungen:

1. Es sei $\gamma \in \mathbb{R}$ mit $\gamma > 0$; für alle $f \in D(M)$ gelte $\{f|f\} \leq \gamma\langle Mf|Mf\rangle$.
2. $v_1,\dots,v_n$ seien linear unabhängige Elemente von $D(M^{-1}N) \cap D(M)$.

3. Es seien $\lambda_* \in \mathbb{R}$ mit $\lambda_* > 0$,
   $A := \gamma(\langle(M-\lambda_*N)v_i \mid (M-\lambda_*N)v_k\rangle)_{i,k=1,\ldots,n}$ und
   $B := (\{v_i \mid v_k\})_{i,k=1,\ldots,n}$.
4. Für $i=1,\ldots,n$ sei μ_i der i-tkleinste Eigenwert der Matrix-Eigenwertaufgabe $Ax = \mu Bx$.

<u>Behauptung:</u> Für alle $r \in \mathbb{N}$ mit $r \leq n$ und $\mu_r < 1$ enthält das Intervall $\left[\frac{\lambda_*}{1+\sqrt{\mu_r}}, \frac{\lambda_*}{1-\sqrt{\mu_r}}\right]$ mindestens r Eigenwerte der Aufgabe $M\phi = \lambda N\phi$.

<u>Beweis:</u> Es sei $r \in \mathbb{N}$ mit $r \leq n$ und $\mu_r < 1$; $\tilde{V}$ sei der Unterraum des $\mathbb{R}^n$, der von denjenigen Eigenvektoren der Aufgabe $Ax = \mu Bx$ aufgespannt wird, die zu einem Eigenwert $\mu_i \leq \mu_r$ gehören. Dann gilt $\dim \tilde{V} \geq r$ und $c'Ac \leq \mu_r c'Bc$ für alle $c \in \tilde{V}$.
$V := \{ \sum_{i=1}^{n} c_i v_i : (c_1,\ldots,c_n)' \in \tilde{V}\}$ ist daher ein Unterraum von $D(M^{-1}N) \cap D(M)$ mit $\dim V \geq r$, und für alle $v \in V$ ergibt sich $\gamma\langle(M-\lambda_*N)v \mid (M-\lambda_*N)v\rangle \leq \mu_r\{v|v\}$.
Hieraus folgt mit Hilfe von (13) für alle $v \in V$ die Abschätzung $\|v - \lambda_* M^{-1}Nv\|^2 \leq \mu_r \|v\|^2$, aus der sich nach Satz 2 die Behauptung ergibt.

Viele der für Randwertaufgaben aufgestellten "expliziten a-priori-Abschätzungen" (s. [52]) können in die Form (13) überführt und dann gemäß Satz 17 zur Einschließung herangezogen werden. So läßt sich die Ungleichung

$$\int_\Omega f^2 dxdy \leq \gamma \int_\Omega (\Delta f)^2 dxdy + \gamma\omega \int_{\partial\Omega} f^2 ds \quad \text{für alle } f \in C^\infty(\bar{\Omega}),$$

in der γ und ω positive, nur von Ω abhängige Zahlen sind, zur Behandlung von Beispiel 3 aus §2 verwenden; hierzu muß man lediglich γ und ω ausrechnen (vgl. [52], S. 22).

Als nächstes wird eine Verallgemeinerung von Satz 2 bewiesen, die es ermöglicht, einige Voraussetzungen von Satz 17 abzuschwächen; hierdurch wird die Einschließung von Eigenwerten Stekloffscher Eigenwertaufgaben wesentlich erleichtert.

<u>Satz 18</u>

<u>Voraussetzungen und Bezeichnungen:</u>

1. $D(N_*)$ sei ein Unterraum von $D(S)$; $N_*: D(N_*) \to H$ sei eine li-

neare Abbildung; es gelte $D(N) \subset D(N_*)$ und $N_*f = Nf$ für alle $f \in D(N)$.

2. Für alle $f,g \in D(N_*)$ gelte $\langle Sf|N_*g\rangle = \langle Sg|N_*f\rangle$ und $\langle Sf|N_*f\rangle \geq 0$. Für alle $f \in D(N_*)$ sei $\|f\|_* := \sqrt{\langle Sf|N_*f\rangle}$.
3. M sei injektiv.
4. Es seien $\varrho, \tau \in \mathbb{R}$ mit $0 \leq \tau < |\varrho|$.
5. V_* sei ein r-dimensionaler Unterraum von $D(M^{-1}N_*) \cap D(N^{-1}N_*)$; für alle $v_* \in V_*$ mit $v_* \neq 0$ gelte $N_*v_* \neq 0$ und
$$\|v_* - \frac{\varrho^2-\tau^2}{\varrho} M^{-1}N_*v_*\|_* \leq \frac{\tau}{|\varrho|}\|v_*\|_* .$$

<u>Behauptung:</u> Im Intervall $[\varrho-\tau, \varrho+\tau]$ liegen mindestens r Eigenwerte der Eigenwertaufgabe $M\phi = \lambda N\phi$.

<u>Beweis</u> (durch Zurückführung auf Satz 2): $V := \{N^{-1}N_*v_*: v_* \in V_*\}$ ist ein r-dimensionaler Unterraum von $D(M^{-1}N)$. Es sei $v \in V$; dann gibt es genau ein $v_* \in V_*$ mit $Nv = N_*v_*$, und es gilt $\|v_*\|_* = \|v\|$ sowie
$$\|v_* - (\varrho^2-\tau^2)\varrho^{-1}M^{-1}N_*v_*\|_* = \|v - (\varrho^2-\tau^2)\varrho^{-1}M^{-1}Nv\| .$$
V erfüllt somit die in Satz 2 genannten Bedingungen.

Satz 17 kann nun folgendermaßen verallgemeinert werden:

<u>Satz 19</u>

<u>Voraussetzungen und Bezeichnungen:</u>

1. Die Voraussetzungen 1 und 2 von Satz 18 seien erfüllt.
2. $D(M_*)$ sei ein Unterraum von $D(N_*)$; $M_*: D(M_*) \to H$ sei eine lineare Abbildung; es gelte $D(M) \subset D(M_*)$ und $M_*f = Mf$ für alle $f \in D(M)$.
3. Es sei $\gamma \in \mathbb{R}$ mit $\gamma > 0$; für alle $f \in D(M_*)$ gelte $\|f\|_*^2 \leq \gamma\langle M_*f|M_*f\rangle$.
4. $v_1,\dots,v_n$ seien Elemente aus $D(M^{-1}N_*) \cap D(N^{-1}N_*) \cap D(M_*)$.
5. Es seien $\lambda_* \in \mathbb{R}$ mit $\lambda_* > 0$,
$A := \gamma(\langle (M_*-\lambda_*N_*)v_i \mid (M_*-\lambda_*N_*)v_k\rangle)_{i,k=1,\dots,n}$ und
$B := (\langle Sv_i|N_*v_k\rangle)_{i,k=1,\dots,n}$; B sei positiv definit.
6. Für $i=1,\dots,n$ sei μ_i der i-tkleinste Eigenwert der Matrix-Eigenwertaufgabe $Ax = \mu Bx$.

<u>Behauptung:</u> Für alle $r \in \mathbb{N}$ mit $r \leq n$ und $\mu_r < 1$ enthält das Inter-

vall $[\frac{\lambda_*}{1+\sqrt{\mu_r}}, \frac{\lambda_*}{1-\sqrt{\mu_r}}]$ mindestens r Eigenwerte der Aufgabe $M\phi = \lambda N\phi$.

Beweis: Es sei $r \in \mathbb{N}$ mit $r \leq n$ und $\mu_r < 1$; $\tilde{V}$ sei der Unterraum des $\mathbb{R}^n$, der von denjenigen Eigenvektoren der Aufgabe $Ax = \mu Bx$ aufgespannt wird, die zu einem Eigenwert $\mu_i \leq \mu_r$ gehören. Dann gilt $\dim \tilde{V} \geq r$ und $c'Ac \leq \mu_r c'Bc$ für alle $c \in \tilde{V}$.
$V_* := \{ \sum_{i=1}^{n} c_i v_i : (c_1,\ldots,c_n)' \in \tilde{V}\}$ ist daher ein Unterraum von $D(M^{-1}N_*) \cap D(N^{-1}N_*) \cap D(M_*)$ mit $\dim V_* \geq r$, und für alle $v_* \in V_*$ mit $v_* \neq 0$ ergibt sich $N_* v_* \neq 0$ sowie
$\gamma\langle (M_*-\lambda_*N_*)v_* \mid (M_*-\lambda_*N_*)v_*\rangle \leq \mu_r\langle Sv_* | N_*v_*\rangle$.
Hieraus folgt mit Hilfe von
$\| v_*-\lambda_*M^{-1}N_*v_*\|_*^2 \leq \gamma\langle M_*v_*-\lambda_*N_*v_* | M_*v_*-\lambda_*N_*v_*\rangle$
für alle $v_* \in V_*$ die Abschätzung
$\| v_*-\lambda_*M^{-1}N_*v_*\|_*^2 \leq \mu_r\| v_*\|_*^2$.
Setzt man $\varrho := \frac{\lambda_*}{1-\mu_r}$ und $\tau := \varrho\sqrt{\mu_r}$, so ergibt sich aus Satz 18 die Behauptung.

Der Beweis von Satz 19 verläuft weitgehend analog zu dem von Satz 17; in mancher Hinsicht weicht er aber auch von dem Schema ab, nach dem die Beweise der anderen Einschließungssätze verlaufen; so tritt beispielsweise die Halbnorm $\|\cdot\|_*$ an die Stelle der Norm $\|\cdot\|$. Natürlich kann man den grundlegenden Satz in §1 so allgemein fassen, daß sich alle Einschließungssätze in genau der gleichen Weise daraus herleiten lassen; man muß dann aber umständlichere Formulierungen in Kauf nehmen.

An Hand eines Beispiels soll nun erläutert werden, wie man Satz 19 auf Stekloffsche Eigenwertaufgaben anwenden kann.

Beispiel 4
Eigenwertaufgabe: $\Delta^2\phi = 0$ in Ω; $\phi = 0$, $\Delta\phi = \lambda\frac{\partial\phi}{\partial n}$ auf $\partial\Omega$
Voraussetzung: γ und ω seien positive Zahlen mit der Eigenschaft, daß
$\int_{\partial\Omega} (\frac{\partial f}{\partial n})^2 ds \leq \gamma\omega\int_{\Omega} (\Delta^2 f)^2 dxdy + \gamma\int_{\partial\Omega} (\Delta f)^2 ds$

für alle $f \in C^{\infty}(\overline{\Omega})$ mit $f = 0$ auf $\partial\Omega$ gilt (bezüglich γ und ω s. [30]).

Einordnung: $H := L_2(\Omega) \times L_2(\partial\Omega)$,

$$\left\langle \begin{pmatrix} f_1 \\ f_2 \end{pmatrix} \middle| \begin{pmatrix} g_1 \\ g_2 \end{pmatrix} \right\rangle := \omega\int_{\Omega} f_1 g_1 dxdy + \int_{\partial\Omega} f_2 g_2 ds \text{ für } \begin{pmatrix} f_1 \\ f_2 \end{pmatrix}, \begin{pmatrix} g_1 \\ g_2 \end{pmatrix} \in H,$$

$$D(M) := D(N) := \{f \in C^{\infty}(\overline{\Omega}) : \Delta^2 f = 0 \text{ in } \Omega,\ f = 0 \text{ auf } \partial\Omega\},$$

$$D(S) := D(M_*) := D(N_*) := \{f \in C^{\infty}(\overline{\Omega}) : f = 0 \text{ auf } \partial\Omega\},$$

$$M_* f := \begin{pmatrix} \Delta^2 f \\ \Delta f_{|\partial\Omega} \end{pmatrix}, \quad M := M_{*|D(M)}, \quad N_* f := Sf := \begin{pmatrix} 0 \\ \frac{\partial f}{\partial n} \end{pmatrix}, \quad N := N_{*|D(N)}.$$

Die zur Anwendung der Sätze 15 und 17 benötigten a-priori-Abschätzungen können im allgemeinen dann leicht aufgestellt werden, wenn (grobe) untere Schranken für den kleinsten Eigenwert gewisser Eigenwertaufgaben bekannt sind (vgl. z. B. [30], [32]).

Die ersten Sätze, in denen zur Eigenwerteinschließung explizite a-priori-Abschätzungen verwendet werden, sind von Kuttler und Sigillito [30], [31], [32] bewiesen worden; in [32] werden Eigenwertaufgaben $M\phi = \lambda\phi$ betrachtet, bei denen M ein symmetrischer Operator ist, während in [30] und [31] Einschließungssätze für zwei Stekloffsche Eigenwertaufgaben zu finden sind. Satz 19 enthält alle diese Resultate als Spezialfälle; er ermöglicht ferner die Behandlung von Eigenwertaufgaben, bei denen der Eigenwert sowohl in der Differentialgleichung als auch in den Randbedingungen auftritt. Während die Sätze von Kuttler und Sigillito nur aussagen, daß in bestimmten Intervallen mindestens e i n Eigenwert der betrachteten Aufgabe liegt, kann man nach Satz 19 auch Intervalle berechnen, die eine größere Mindestanzahl von Eigenwerten enthalten; dies dürfte für die Behandlung von Eigenwertaufgaben mit mehrfachen oder sehr nahe beieinander liegenden Eigenwerten von Interesse sein. - Erwähnt seien noch die Ergebnisse von McLaurin [40], [41], die als Vorläufer der Sätze von Kuttler und Sigillito aufgefaßt und ebenfalls aus dem grundlegenden Satz in §1 hergeleitet werden können. - Kuttler und Sigillito haben ihre Sätze mit gutem Erfolg an zwei Stekloffschen Eigenwertaufgaben [30], [31] und zahlreichen Schwingungsproblemen (Berechnung von Eigenfrequenzen rhombischer Membranen [32], orthotroper Rechteckplatten [34] sowie dreieckiger, trapezförmiger und rhombischer Platten [33], [35]) numerisch erprobt.

Die Tatsache, daß sich so viele bekannte Ergebnisse in einheitlicher Weise herleiten lassen, legt die Vermutung nahe, daß die geschilderte Methode auch zur Aufstellung neuer Einschließungssätze beitragen kann.

Die Verfasser danken der Deutschen Forschungsgemeinschaft für die großzügige Förderung und Herrn Dr. H.-E. Lahmann für die freundliche Unterstützung ihres Forschungsvorhabens.

Literatur

[1] Albrecht, J.: Verallgemeinerung eines Einschließungssatzes von L. Collatz. Z. Angew. Math. Mech. 48 (1968), T43 - T46

[2] Albrecht, J.: Quotienteneinschließungssätze für allgemeine Eigenwertaufgaben. Wiss. Z. Tech. Univ. Dresden 29 (1980), 424 - 426

[3] Albrecht, J.: Einschließung von Eigenwerten bei Schwingungen von Kreisbögen. Z. Angew. Math. Mech. 63 (1983), 387 - 389

[4] Andrushkiw, R.I.: On the approximate solution of K-positive eigenvalue problems Tu - λSu = O. J. Math. Anal. Appl. 50 (1975), 511 - 529

[5] Aubin, J.-P.: Approximation of elliptic boundary-value problems. New York - London - Sydney - Toronto: Wiley-Interscience 1972

[6] Bartsch, H.: Ein Einschließungssatz für die charakteristischen Zahlen allgemeiner Matrizen-Eigenwertaufgaben. Arch. Math. (Basel) 4 (1953), 133 - 136

[7] Bauer, F.L. und A.S. Householder: Moments and characteristic roots. Numer. Math. 2 (1960), 42 - 53

[8] Bazley, N.W. und D.W. Fox: Truncations in the method of intermediate problems for lower bounds to eigenvalues. J. Res. Nat. Bur. Standards Sect. B 65 (1961), 105 - 111

[9] Bredehöft, M. und W. Hauger: An iteration procedure and bounds for the eigenvalues for a class of non-selfadjoint eigenvalue problems. Mech. Res. Comm. 6 (1979), 105 - 111

[10] Collatz, L.: Einschließungssatz für die charakteristischen Zahlen von Matrizen. Math. Z. 48 (1942), 221 - 226

[11] Collatz, L.: Eigenwertaufgaben mit technischen Anwendungen. 2. Auflage. Leipzig: Akademische Verlagsgesellschaft Geest & Portig KG 1963

[12] Collatz, L.: Einschließungssatz für Eigenwerte bei partiellen Differentialgleichungen 2. und 4. Ordnung. Z. Angew. Math. Mech. 43 (1963), 277 - 280

[13] Dettmar, H.-K.: Symmetrisierbare Eigenwertaufgaben bei gewöhnlichen linearen Differentialgleichungen. Z. Angew. Math. Mech. 34 (1954), 284 - 287

[14] Donnelly, J.D.P.: Eigenvalues of membranes with reentrant corners. SIAM J. Numer. Anal. 6 (1969), 47 - 61

[15] Falk, S.: Einschließungssätze für die Eigenwerte normaler Matrizenpaare. Z. Angew. Math. Mech. 44 (1964), 41 - 55

[16] Fichera, G.: Numerical and quantitative analysis. London - San Francisco - Melbourne: Pitman 1978

[17] Fox, L., P. Henrici und C. Moler: Approximations and bounds for eigenvalues of elliptic operators. SIAM J. Numer. Anal. 4 (1967), 89 - 102

[18] Goerisch, F.: Über Quotienten-Einschließungssätze bei allgemeinen Eigenwertaufgaben. In: J. Albrecht, L. Collatz und G. Hämmerlin (Hrsg.): Numerische Behandlung von Differentialgleichungen mit besonderer Berücksichtigung freier Randwertaufgaben. International Series of Numerical Mathematics (ISNM), Vol. 39. Basel - Stuttgart: Birkhäuser 1978, 86 - 100

[19] Goerisch, F.: Weiterentwicklung von Verfahren zur Berechnung von Eigenwertschranken. Dissertation TU Clausthal 1978

[20] Goerisch, F.: Eine Verallgemeinerung eines Verfahrens von N.J. Lehmann zur Einschließung von Eigenwerten. Wiss. Z. Tech. Univ. Dresden 29 (1980), 429 - 431

[21] Goerisch, F.: Über die Anwendung einer Verallgemeinerung des Lehmann-Maehly-Verfahrens zur Berechnung von Eigenwertschranken. In: J. Albrecht und L. Collatz (Hrsg.): Numerische Behandlung von Differentialgleichungen, Band 3. International Series of Numerical Mathematics (ISNM) Vol. 56. Basel - Boston - Stuttgart: Birkhäuser 1981, 58 - 72

[22] Goerisch, F.: Ein Stufenverfahren zur Berechnung von Eigenwertschranken. Erscheint demnächst in den Nova Acta Leopoldina (N.F.)

[23] Goerisch, F. und J. Albrecht: Untere Schranken für die Eigenwerte Stekloffscher Eigenwertaufgaben. In: M. Gregus (Hrsg.): Equadiff 5. Proceedings. Teubner-Texte zur Mathematik, Bd. 47. Leipzig: Teubner 1982, 9 - 13

[24] Goerisch, F. und J. Albrecht: Die Monotonie der Templeschen Quotienten. Eingereicht bei der Z. Angew. Math. Mech.

[25] Goerisch, F. und H. Haunhorst: Eigenwertschranken für Eigenwertaufgaben mit partiellen Differentialgleichungen. Eingereicht bei der Z. Angew. Math. Mech.

[26] Hadeler, K.P.: Einschließungssätze bei normalen und bei positiven Operatoren. Arch. Rational Mech. Anal. 21 (1966), 58 - 88

Harazov, D.F. siehe Kharazov, D.F.

[27] Held, W.: Einschließungen von Eigenwerten. Dissertation TU Clausthal 1972

[28] Held, W.: Die Collatzschen Einschließungssätze für Eigenwerte bei Differentialgleichungen. In: L. Collatz und K.P. Hadeler (Hrsg.): Numerische Behandlung von Eigenwertaufgaben. International Series of Numerical Mathematics (ISNM), Vol. 24. Basel - Stuttgart: Birkhäuser 1974, 47 - 55

[29] Kharazov, D.F.: Estimates for the eigenvalues of certain operators with discrete spectrum. Differencial'nye Uravnenia 1 (1965), 1054 - 1069 (Englische Übersetzung: Differential Equations 1 (1965), 822 - 834)

[30] Kuttler, J.R.: Dirichlet eigenvalues. SIAM J. Numer. Anal. 16 (1979), 332 - 338

[31] Kuttler, J.R.: Bounds for Stekloff eigenvalues. SIAM J. Numer. Anal. 19 (1982), 121 - 125

[32] Kuttler, J.R. und V.G. Sigillito: Bounding eigenvalues of elliptic operators. SIAM J. Math. Anal. 9 (1978), 768 - 773

[33] Kuttler, J.R. und V.G. Sigillito: Upper and lower bounds for frequencies of clamped rhombical plates. Journal of Sound and Vibration 68 (1980), 597 - 607

[34] Kuttler, J.R. and V.G. Sigillito: Upper and lower bounds for the frequencies of clamped orthotropic plates. Journal of Sound and Vibration 73 (1980), 247 - 259

[35] Kuttler, J.R. and V.G. Sigillito: Upper and lower bounds for frequencies of trapezoidal and triangular plates. Journal of Sound and Vibration 78 (1981), 585 - 590

[36] Lange, E., Diplomarbeit TU Clausthal 1979

[37] Lehmann, N.J.: Beiträge zur numerischen Lösung linearer Eigenwertprobleme. I. Z. Angew. Math. Mech. 29 (1949), 341 - 356

[38] Lehmann, N.J.: Beiträge zur numerischen Lösung linearer Eigenwertprobleme. II. Z. Angew. Math. Mech. 30 (1950), 1 - 16

[39] Lehmann, N.J.: Optimale Eigenwerteinschließungen. Numer. Math. 5 (1963), 246 - 272

[40] McLaurin, J.: Bounding eigenvalues of clamped plates. Z. Angew. Math. Phys. 19 (1968), 676 - 681

[41] McLaurin, J.: Bounds for vibration frequencies and buckling loads of clamped plates. Dissertation ETH Zürich 1969

[42] Moler, C.B. und L.E. Payne: Bounds for eigenvalues and eigenvectors of symmetric operators. SIAM J. Numer. Anal. 5 (1968), 64 - 70

[43] Nickel, K.L.E.: Extension of a recent paper by Fox, Henrici and Moler on eigenvalues of elliptic operators. SIAM J. Numer. Anal. 4 (1967), 483 - 488

[44] Rektorys, K.: Variational methods in mathematics, science and engineering. Dordrecht - Boston: D. Reidel Publishing Company 1977

[45] Rektorys, K. und Vospěl, Z.: On a method of twosided eigenvalue estimates for elliptic equations of the form $Au - \lambda Bu = 0$. Apl. Mat. 26 (1981), 211 - 240

[46] Rieder, G.: Elementargeometrische Herleitung von Einschließungssätzen für Eigenwerte. Z. Angew. Math. Mech. 48 (1968), 207 - 210

[47] Ruppert, K.: Über die Anwendung von Quotienten-Einschließungssätzen für Eigenwerte. Dissertation TU Clausthal 1983

[48] Schäfke, F.W. und A. Schneider: S-hermitesche Rand-Eigenwertprobleme. I. Math. Ann. 162 (1965), 9 - 26

[49] Schäfke, F.W. und A. Schneider: S-hermitesche Rand-Eigenwertprobleme. II. Math. Ann. 165 (1966), 236 - 260

[50] Schellhaas, H.: Verfahren zur expliziten Berechnung unterer Eigenwertschranken beim allgemeinen Eigenwertproblem im Hilbertraum. Dissertation TH Darmstadt 1966

[51] Schellhaas, H.: Ein Verfahren zur Berechnung von Eigenwertschranken mit Anwendung auf das Beulen von Rechteckplatten. Ingenieur-Archiv 37 (1968), 243 - 250

[52] Sigillito, V.G.: Explicit a priori inequalities with applications to boundary value problems. London - San Francisco - Melbourne: Pitman 1977

[53] Stummel, F.: Rand- und Eigenwertaufgaben in Sobolewschen Räumen. Berlin - Heidelberg - New York: Springer 1969

[54] Temple, G.: The theory of Rayleigh's principle as applied to continuous systems. Proc. Roy. Soc. Ser. A 119 (1928), 276 - 293

[55] Velte, W.: Direkte Methoden der Variationsrechnung. Stuttgart: Teubner 1976

[56] Weinstein, A. und W. Stenger: Methods of intermediate problems for eigenvalues. New York - London: Academic Press 1972

[57] Wielandt, H.: Ein Einschließungssatz für charakteristische Wurzeln normaler Matrizen. Arch. Math. (Basel) 1 (1949), 348 - 352

Übersicht über die Herleitung der Einschließungssätze

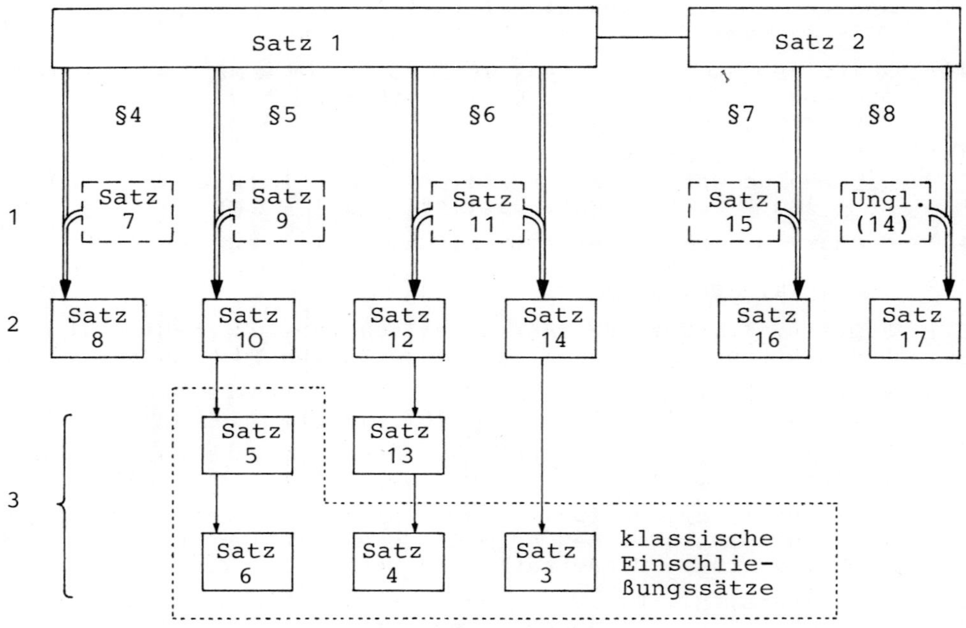

1 Normabschätzungen 2 Einschließungssätze 3 Spezialfälle

Dr. F. Goerisch, Prof. Dr. J. Albrecht
Institut für Mathematik der TU Clausthal
Erzstraße 1, D 3392 Clausthal-Zellerfeld

ISNM, Vol. 69
Numerical Treatment of
Eigenvalue Problems, Vol. 3

PROGRESS ON ESTIMATION OF ENERGY LEVELS FOR MULTI-ELECTRON ATOMS*

W. M. Greenlee and D. M. Russell

A new base problem for the calculation of lower bounds to the energy levels of multi-electron atoms is presented in this paper. The effective field method is used to produce a central field approximation whose energy levels are lower bounds to those of the full Hamiltonian. In contrast to previous calculations with the effective field method, the lower bound Hamiltonian given here is explicitly resolvable, rather than only numerically solvable. This is accomplished by employing the Hulthen potential, and results in a base problem which is suitable as the starting point for the method of intermediate problems to improve the lower bounds. Lower bounds for the ground states of the neutral atoms with three to thirty electrons are presented.

1. INTRODUCTION.

We consider the classical fixed nucleus, spin free Hamiltonian for an N-electron atom with nuclear charge Z. In atomic units this may be written

$$H = \sum_{i=1}^{N} \left(-\frac{1}{2}\Delta_i - \frac{Z}{r_i}\right) + \sum_{i<j}^{N} \frac{1}{r_{ij}}, \qquad (1)$$

where $r_i = |\vec{x}_i|$, $\vec{x}_i \in \mathbb{R}^3$, is the distance of the ith electron from

*This research was partially supported by NSF Grants 02MCS-7902663 and MCS-8301402.

the nucleus, $r_{ij} = |\vec{x}_i - \vec{x}_j|$, and Δ_i is the Laplacian operating on the ith copy of $\mathbb{R}^3$. In these units Z is a positive integer, and for neutral atoms, $N = Z$. H is essentially self-adjoint on the subspace $C_0^\infty(\mathbb{R}^{3N})$ of $L^2(\mathbb{R}^{3N})$, and we retain the symbol H for the unique self adjoint extension. H is bounded from below with discrete lower spectrum (cf. [18]).

The energy levels of the atom or ion modeled by 1) are the eigenvalues of H restricted to subspaces determined by the Pauli exclusion principle. It is these eigenvalues that we wish to approximate. The Pauli principle is an empirically based physical law first enunciated by Pauli in 1925 on the basis of experimental evidence. It restricts the domain of the Hamiltonian to subspaces with certain symmetry properties, thereby coupling the electron spin to the electrostatic term. Thus we wish to compute spectral approximations to the reduced operator corresponding to physically realizable states. For atoms and positive ions, i.e., $N \leq Z$, the resulting physical Hamiltonians are also bounded from below with discrete lower spectrum (cf. [3,22,25]). Negative ions, i.e., $N > Z$, have at most finitely many bound states (eigenvalues) below the continuum - possibly none at all (cf. [17,24,26]). Rather than attempting to explicitly use the complicated reduced operator, one works with H and takes steps to ensure that the required symmetry properties of the approximate eigenvectors are satisfied.

Extensive upper bound calculations for energy levels have been carried out by the Hartree-Fock method (cf. [7,14]). On the other hand, accurate lower bounds have proved to be quite elusive.

In the case of hydrogen, i.e., $N = Z = 1$, the eigenvalue problem for H is explicitly solvable. This and separation of variables leads to solvability of the eigenvalue problem for

$$H_0 = \sum_{i=1}^{N} \left(-\frac{1}{2}\Delta_i - \frac{Z}{r_i}\right).$$

The latter yields crude lower bounds to the eigenvalues of H since the operator

$$H_1 = \sum_{i<j}^{N} \frac{1}{r_{ij}}$$

is positive.

The first major breakthrough in lower bounds to atomic eigenvalues was due to Bazley [4]. Highly accurate lower bounds for the S states of parahelium, $N = Z = 2$, were obtained by introducing the method of special choice, the first numerically workable variant of Aronszajn's method [2]. But for lithium, $N = Z = 3$, accurate rigorous lower bounds are still unknown. The ground state (lowest energy level) of lithium lies above some of the essential spectrum of the base problem obtained from H_0. Stability of the essential spectrum to compact perturbations thus makes accurate approximation of the energy levels impossible by the finite rank methods of Aronszajn type. This part of the problem was solved by Fox [10], who introduced numerically workable perturbations of infinite rank into atomic problems and thus made possible the movement of essential spectra. This technique was used by Fox and Sigillito [11,12] to obtain accurate lower bounds to the lower energy levels of radial lithium, a somewhat simplified model for the lithium Hamiltonian which retained the difficulty with the essential spectrum. Accurate lower bounds for the Li^+ ion $(N = 2, Z = 3)$ were obtained by Freund [13], who also showed that Fox's method moves the essential spectrum enough to make possible the calculation of good lower bounds for lithium.

Part of the difficulty with the above approach is due to the fact that in the above decomposition of H, H_1 is a rather large perturbation of H_0, particularly in the case of large atoms. We have chosen to seek another decomposition of H,

$$H = H^o + H',$$

with H^o resolvable and $H' > 0$ as before, and such that H^o has higher eigenvalues than H_0, while the eigenvectors of H^o are at least as tractable as those formed from the traditional Coulomb wave functions. To accomplish this, we make use of the effective field method [15]. In [15,16,23] the effective field method has been used to calculate lower bounds from choices for H^o which are only amenable to computer solution.

This paper outlines the construction of a base problem H^o with the desirable properties mentioned above. The full details form a portion of [21]. In the next section we sketch the effective field method. Then in Section 3, the Hulthén potential is used to construct an explicitly resolvable lower bound operator H^o. H^o is then suitable as a base operator for Fox's method to improve the lower bounds. The resulting intermediate problems are discussed in [21], with particular emphasis on the beryllium atom, $N = Z = 4$. In Section 4 we record some remarks, and provide base problem numerical results for the neutral atoms having three to thirty electrons.

2. THE EFFECTIVE FIELD METHOD.

As indicated in the introduction, a major difficulty with the traditional decomposition of H is that the interelectron repulsion term H_1 is quite a large perturbation of H_0. Considerable progress toward a less drastic perturbation can be made via the assumption that the interelectron repulsion term contains a large spherically symmetric component. That this assumption has merit is borne out by upper bound results from the Hartree-Fock method (cf. [14]). Thus we wish to construct a spherically symmetric, single particle potential energy function $V(r_i)$, which is meant to approximate the actual potential energy of the ith electron by an averaged field due to the nucleus and the other $N - 1$ electrons. Then we may write $H = H^o + H'$ with

$$H^o = \sum_{i=1}^{N} \left(-\frac{1}{2}\Delta_i + V(r_i)\right),$$

and

$$H' = \sum_{i<j}^{N} \frac{1}{r_{ij}} - \sum_{i=1}^{N} \frac{Z}{r_i} - \sum_{i=1}^{N} V(r_i).$$

It follows that H^o will be explicitly resolvable by separation of variables if $V(r_i)$ is chosen so that the resulting radial equations are explicitly solvable. We also require that H' be nonnegative so as to obtain a lower bound Hamiltonian. To accomplish this, we appeal to the effective field theorem of [16].

Recall that a bounded, complex valued, continuous function f on $\mathbb{R}^n$ is called a _function of positive type_ if $[f(\vec{x}_i - \vec{x}_j)]_{i,j}$ is a positive matrix on $\mathbb{C}^M$ for each M and all $\vec{x}_1,\ldots,\vec{x}_M \in \mathbb{R}^n$. These functions form a cone, and Bochner's theorem states that the cone of functions of positive type is exactly the set of Fourier transforms of the finite, positive measures on $\mathbb{R}^n$ (cf. [20]). We now state the effective field theorem as originally given in [16]. The different version of this theorem presented later in [23] does not apply to our choice of potential.

EFFECTIVE FIELD THEOREM. _Let_ $v(\vec{x})$, $\vec{x} \in \mathbb{R}^3$, _be a real function of positive type, and let the real function_ $\Phi(\vec{x})$ _be the Fourier transform of a real integrable function. Then,_

$$\sum_{i=1}^{N} \Phi(\vec{x}_i) - \frac{1}{2} \int \frac{d^3k}{(2\pi)^3} \left[\frac{\tilde{\Phi}^2(\vec{k})}{\tilde{v}(\vec{k})} + N\,\tilde{v}(\vec{k})\right] \leq \sum_{i<j}^{N} v(\vec{x}_i - \vec{x}_j),$$

where "~" _denotes the Fourier transform,_ $\vec{k}$ _the transform variable, and_ d^3k _the Lebesgue measure on_ $\mathbb{R}^3$.

If the potential $\vec{v}(x)$ of positive type is now chosen so that

$$v(\vec{x}) \leq \frac{1}{r} = \frac{1}{|\vec{x}|},$$

we then obtain the _effective field inequality_,

$$\sum_{i=1}^{N} \Phi(\vec{x}_i) - N\Lambda < \sum_{i<j}^{N} \frac{1}{r_{ij}},$$

where Λ is the positive constant:

$$\Lambda = \frac{1}{2N} \int \frac{d^3k}{(2\pi)^3} \left[\frac{\tilde{\Phi}^2(\vec{k})}{\tilde{v}(\vec{k})} + N\,\tilde{v}(\vec{k})\right].$$

The potentials $v(\vec{x})$ and $\Phi(\vec{x})$ are called the <u>subsidiary potential</u> and the <u>repulsion potential</u>, respectively.

Following [16], a useful choice for the subsidiary potential is

2) $$v(\vec{x}) = \frac{1 - e^{-\mu|\vec{x}|}}{|\vec{x}|},$$

where μ is a positive parameter. The Fourier transform of this potential is

$$\tilde{v}(\vec{k}) = \frac{4\pi\mu^2}{k^2(k^2 + \mu^2)},$$

where $k = |\vec{k}|$. A repulsion potential that should be physically reasonable is

$$\Phi(\vec{x}) = \int d^3x' \, \frac{\rho(\vec{x}')}{|\vec{x} - \vec{x}'|},$$

where $\rho(\vec{x})$ is the electron density function. So $\Phi(\vec{x})$ satifies Poisson's equation with charge density $-\rho(\vec{x})$:

$$\Delta\Phi(\vec{x}) = -4\pi\rho(\vec{x}).$$

By the convolution theorem,

$$\tilde{\Phi}(\vec{k}) = \frac{4\pi\tilde{\rho}(\vec{k})}{k^2}.$$

With these choices of $v(\vec{x})$ and $\Phi(\vec{x})$, it follows with the use of $v(\vec{0}) = \mu$, Parseval's relation, and Fubini's theorem, that

$$3)\qquad \Lambda = \frac{1}{2N}\left\{ \frac{4\pi}{\mu^2}\int|\rho(\vec{x})|^2 d^3x + \int \Phi(\vec{x})\rho(\vec{x})d^3x + \mu N \right\}.$$

It is immediate that the optimal subsidiary potential of the form 2) is now obtained by minimizing Λ with respect to μ, and the minimum is attained for

$$\mu = \left[\frac{8\pi}{N}\int d^3x|\rho(\vec{x})|^2\right]^{1/3} .$$

We now specialize to a radially symmetric repulsion potential $\Phi(r)$, where we use a standard abuse of notation. By letting

$$\chi(r) = \frac{r}{Z}\Phi(r),$$

it follows that

$$\Delta\Phi = \frac{1}{r}\frac{d^2}{dr^2}(r\Phi) = \frac{Z}{r}\frac{d^2\chi}{dr^2} .$$

Then by use of the addition theorem for spherical harmonics, 3) can be rewritten in terms of χ,

$$4)\qquad \Lambda = \frac{1}{2N}\left\{Z^2\int_0^\infty \left(\frac{d\chi}{dr}\right)^2 dr + \frac{Z^2}{\mu^2}\int_0^\infty \left(\frac{d^2\chi}{dr^2}\right)^2 dr + N\mu\right\},$$

with minimizing μ given by,

$$5)\qquad \mu = (2Z)^{1/3}\left[\int_0^\infty \left(\frac{d^2\chi}{dr^2}\right)^2 dr\right]^{1/3} .$$

3. THE NEW BASE PROBLEM.

We now rewrite 1) as $H = H^o + H'$ with

$$H^o = \sum_{i=1}^{N}\left[-\frac{1}{2}\Delta_i - \frac{Z}{r_i} + \Phi(r_i) - \Lambda\right],$$

and

$$H' = \sum_{i<j}^{N} \frac{1}{r_{ij}} + (N\Lambda - \sum_{i=1}^{N} \Phi(r_i)),$$

where Φ and Λ are as above. The effective field theorem implies that $H' \geq 0$, so the eigenvalues of H^{0} furnish lower bounds to those of H. The spectral problem for H^{0} will be explicitly solvable if that is true of the coresponding radial problems,

$$[-\frac{1}{2}(\frac{1}{r_i^{2}}\frac{d}{dr_i} r_i^{2}\frac{d}{dr_i}) - \frac{Z}{r_i} + \Phi(r_i) + \frac{\ell(\ell+1)}{2r_i^{2}} - \Lambda]\phi = \varepsilon\phi,$$

in $L^{2}(\mathbb{R}^{+}, r_i^{2}dr_i)$ for $\ell = 0,1,2,\ldots$, and $i = 1,2,\ldots,N$. Letting $\bar{\varepsilon} = \varepsilon + \Lambda$ and suppressing the subscript i, the radial eigenvalue problems take the form

6) $$[-\frac{1}{2}(\frac{1}{r^{2}}\frac{d}{dr} r^{2}\frac{d}{dr}) - \frac{Z}{r} + \Phi(r) + \frac{\ell(\ell+1)}{2r^{2}}]\phi = \bar{\varepsilon}\,\phi, \quad \ell = 0,1,2,\ldots .$$

Physical arguments based on the idea of screening (cf. [6]) indicate that the potential energy function

$$V(r) = \Phi(r) - \frac{Z}{r}$$

in our central field approximation should be chosen so that

$$V(r) = \begin{cases} -\frac{Z}{r} + C + O(r), & \text{as } r \to 0, \\ o(1), & \text{as } r \to \infty, \end{cases}$$

with $C > 0$. Herein, O and o are the usual Landau symbols. We know of no potentials satisfying these conditions for which 6) is explicitly solvable for all ℓ. Moreover, Φ must satisfy the hypotheses of the effective field theorem.

However, additional latitude is obtained by only seeking exact solutions to the S-wave, i.e., $\ell = 0$, equation. Since for $\ell = 1,2,\ldots$, $\ell(\ell+1)/2r^{2}$ is a positive multipliction operator, we may obtain lower bounds from the S-wave equation and partially restore the effect of nonzero ℓ by use of the method of intermediate problems on these radial equations. Since the quantity $\ell(\ell + 1)$ arises as a separation constant, it is not entirely obvious that this will yield a new base problem for H. This is however the case, as is shown with the use of abstract separation of variables in [21]. Special choice is used in these radial intermediate problems since it is not only easily carried out, but the intermediate radial eigenvectors are easily obtained from those of the S-wave equation. Thus our new base problem is explicitly solvable provided that the S-wave equation is.

A potential meeting the above physical criteria for which the S-wave equation is explicitly solvable with tractable eigenfunctions is the Hulthén potential

$$V(r) = -\frac{Z}{a}\frac{e^{-r/a}}{1 - e^{-r/a}} ,$$

where a is a positive scaling parameter. Thus we have that:

$$\Phi(r) = \frac{Z}{r} - \frac{Z}{a}\frac{e^{-r/a}}{1 - e^{-r/a}} .$$

The Hulthén potential has previously been used as an approximate atomic potential by scattering theorists (cf. [19]).

That Φ also satisfies the conditions of the effective field theorem can be shown as follows. First of all, recall that $\tilde{\Phi}(\vec{k}) = 4\pi\tilde{\rho}(\vec{k})/k^2$. So if $\tilde{\rho}$ is a continuous function in $L^2(\mathbb{R}^3)$, we find that

$$\int |\tilde{\Phi}(\vec{k})| d^3k = 4\pi \int_{|\vec{k}|<1} \frac{\tilde{\rho}(\vec{k})}{k^2} d^3k + 4\pi \int_{|\vec{k}|>1} \frac{\tilde{\rho}(\vec{k})}{k^2} d^3k < \infty .$$

The first integral on the right is finite since $\tilde{\rho}$ is continuous, while the Cauchy-Schwarz inequality shows that the second is finite since $\tilde{\rho} \in L^2(\mathbb{R}^3)$. Now

$$-4\pi\rho(\vec{x}) = \frac{Z}{|\vec{x}|}\chi''(|\vec{x}|), \quad ' = d/dr,$$

and $\chi''(r)$ is a smooth function on $[0, \infty)$ which decays exponentially as $r \to \infty$. It follows immediately that $\rho(\vec{x}) \in L^1(\mathbb{R}^3) \cap L^2(\mathbb{R}^3)$, and thus $\tilde{\rho}$ is a continuous function in $L^2(\mathbb{R}^3)$ which also vanishes at infinity. So the Fourier inversion formula shows that Φ is the Fourier transform of an integrable function. Since Φ is also real and radially symmetric, it is now a consequence of the integral representation of Φ as an inverse Fourier transform that $\tilde{\Phi}$ is also real.

The eigenvalue problem 6) with $\ell = 0$ and $\Phi(r) - Z/r$ given by the Hulthén potential can be written in the form,

7) $$-\frac{1}{2}\left[\frac{1}{r}\frac{d^2}{dr^2}(r\phi)\right] - \frac{Z}{a}\frac{e^{-r/a}}{1-e^{-r/a}}\phi = \bar{\varepsilon}\phi,$$

with the conditions $\phi \in L^2((0,\infty), r^2dr)$ and $\phi(r) = o(1)$ as $r \to \infty$. The solution of this problem is given in [8]. Briefly, by letting $\bar{\phi}(r) = r\phi(r)$, $x = r/a$, $z = e^{-x}$, and setting

$$\alpha^2 = -2\,\bar{\varepsilon}\,a^2 > 0 \quad , \quad \beta^2 = 2\,Z\,a > 0,$$

one gets (with a standard abuse of notation) the equation,

8) $$z^2\frac{d^2\bar{\phi}}{dz^2} + z\frac{d\bar{\phi}}{dz} + \left(-\alpha^2 + \beta^2\frac{z}{1-z}\right)\bar{\phi} = 0,$$

subject to the boundary conditions,

9) $$\bar{\phi}(z = 0) = \bar{\phi}(z = 1) = 0.$$

The substitution $\bar{\phi} = z^{\alpha}(1-z)w(z)$ now transforms 8) into the hypergeometric equation,

$$z(1-z)w'' + \left[(2\alpha+1) - (2\alpha+3)z\right]w' - (2\alpha+1-\beta^2)w = 0,$$

which has the general solution,

$$w(z) = A\ {}_2F_1(\alpha + 1 + \gamma,\ \alpha + 1 - \gamma,\ 2\alpha + 1;\ z)$$
$$+ B\ z^{-2\alpha}\ {}_2F_1(-\alpha + 1 + \gamma,\ -\alpha + 1 - \gamma,\ -2\alpha + 1;\ z),$$

with $\gamma = (\alpha^2 + \beta^2)^{1/2}$. The first of the boundary conditions 9) now implies that $B = 0$. Then by use of identities for the ${}_2F_1$ functions (cf. [1,8]), it can be shown that the second of the boundary conditions 9) requires that

$$\text{10)} \qquad \alpha - \gamma = -n,\ n = 1,2,3,\ldots,$$

where α denotes the positive square root of α^2.

The corresponding (unnormalized) solution of 7) is now,

$$\phi(r) = e^{-\alpha r/a}\left(\frac{1 - e^{-r/a}}{r}\right)\ {}_2F_1(2\alpha+1+n,\ 1-n,\ 2\alpha+1;\ e^{-r/a}).$$

Since ${}_2F_1(2\alpha+1+n,\ 1-n,\ 2\alpha+1;\ e^{-r/a})$ is a polynomial in $e^{-r/a}$, the eigenfunctions of our base problem are certainly as tractable as could be hoped for. The condition 10) yields $\alpha = (\beta^2 - n^2)/2n$, and since $\alpha > 0$, we must have $\beta^2 > n^2$ in order that eigenvalues occur. This determines a finite number of eigenvalues corresponding to a given value of the scaling parameter a. The nth eigenvalue $\bar{\varepsilon}_n$ is now easily seen to be

$$\bar{\varepsilon}_n = -\frac{Z}{a}\left[\frac{\beta^2 - n^2}{2n\beta}\right]^2 = -\frac{Z}{a}\left[\frac{2Za - n^2}{2n\beta}\right]^2,\ n = 1,2,\ldots,(2Za)^*,$$

where $(2Za)^*$ denotes the greatest integer less than $2Za$.

The Hulthén potential is a decreasing function of a for each r, dominates the Coulomb potential, and the difference of the two tends uniformly to zero on $[0, \infty)$ as $a \to \infty$. These facts are reflected in the inequality

$$\bar{\varepsilon}_n = -\frac{Z^2}{2n^2} + \frac{Z}{2a} - \frac{n^2}{8a^2} > -\frac{Z^2}{2n^2}.$$

The latter term is the nth eigenvalue of the Coulombic radial equation, and the inequality holds since $2Za > n^2$. Recalling that $\bar{\varepsilon} = \varepsilon + \Lambda$, the eigenvalues of the original radial problem are,

$$\varepsilon_n = \bar{\varepsilon}_n - \Lambda = -\frac{Z}{a}\left(\frac{2Za - n^2}{2n\beta}\right)^2 - \Lambda\ , \quad n = 1,2,\ldots,(2Za)^*,$$

where Λ is given by 4), 5). A tedious but direct calculation now yields,

$$\int_0^\infty \left(\frac{d\chi}{dr}\right)^2 dr = \frac{1}{a}\left(\frac{5}{6} - \frac{\pi^2}{18}\right), \quad \int_0^\infty \left(\frac{d^2\chi}{dr^2}\right)^2 dr = \frac{1}{a^3}\left(\frac{1}{6} - \frac{\pi^2}{90}\right),$$

so that $\Lambda = Z\,\lambda/2a$ with

$$\lambda = \left(\frac{5}{6} - \frac{\pi^2}{18}\right) + 3(2Z)^{-2/3}\left(\frac{1}{6} - \frac{\pi^2}{90}\right)^{1/3},$$

and

$$\varepsilon_n = -\frac{Z^2}{2n^2} - \frac{n^2}{8a^2} + \frac{Z}{2a}(1 - \lambda),\ n = 1,2,\ldots,(2Za)^*.$$

The parameter a now assumes an important role. It is adjusted so that the Pauli exclusion principle is satisfied in an optimal, or at least near optimal way. Briefly, the potential well must be deep enough, i.e., a must be large enough, that enough states (eigenvectors) exist so that the Pauli principle can be satisfied. Yet it is undesirable to take a much larger than necessary, since then ε_n decreases toward the Coulombic value. Appropriate methods by which to choose a are given in [21].

4. REMARKS AND NUMERICAL RESULTS.

Recall that the base problem is formed by solving S-wave radial equations and partially restoring the nonzero ℓ contributions by special choice. With the finite number of eigenvectors available for special choice, no serious attempt has been made to obtain with precision the eigenvalues of the "unsolvable base problem" obtained directly from the effective field method. The rationale for not continuing further with base problem calculations is twofold: the limited use of special choice already provides significant movement, and the corresponding eigenfunctions are easily obtained and simple enough to be feasible to use in the intermediate problems of Fox's method. A convergence theory for these intermediate problems is given in [5].

There is another advantage of the base problem described in the previous section which is in part due to the finite number of eigenvalues. The hydrogenic (Coulombic) base problem with an infinity of eigenvalues has the so-called "accidental degeneracy problem", namely eigenvalues which are degenerate with respect to the quantum number ℓ. This makes the resolution of intermediate problem spectra exceedingly difficult, and has long plagued upper bound methods as well.

The base problem of Section 3 was used in [21] to calculate lower bounds to the ground states of the neutral atoms from $Z = 3$ to $Z = 30$. We give these results in the base problem column of the following table. The upper bound column is nonrelativistic Hartree-Fock data from [7], while the numbers in the previous effective field column are from the effective field calculations reported in [23]. The Thomas-Fermi values are approximations to the energy levels obtained by the Thomas-Fermi statistical method, which is neither a lower bound nor an upper bound method. This data comes from equations given in [9]. The right hand column records the lower bounds given by the traditional hydrogenic base problem. The tabulated data is only given to three significant figures since that is the accuracy given in [23]. Much greater accuracy is of course required to use the base problem of Section 3 in the method of intermediate problems. Such accuracy is obtained in [21].

GROUND STATE ATOMIC ENERGY TABLE (IN ATOMIC UNITS)*

Z	BASE PROBLEM	UPPER BOUND	PREVIOUS EFFECTIVE FIELD	THOMAS-FERMI	HYDROGENIC
3	-9.22 00	-7.43 00		-9.98 00	-1.01 01
4	-1.77 01	-1.46 01		-1.95 01	-2.00 01
5	-3.06 01	-2.45 01		-3.29 01	-3.44 01
6	-4.76 01	-3.77 01		-5.03 01	-5.40 01
7	-6.96 01	-5.43 01		-7.21 01	-7.96 01
8	-9.71 01	-7.48 01		-9.84 01	-1.12 02
9	-1.31 02	-9.94 01		-1.30 02	-1.52 02
10	-1.71 02	-1.29 02	-1.64 02	-1.66 02	-2.00 02
11	-2.13 02	-1.62 02		-2.07 02	-2.49 02
12	-2.59 02	-2.00 02		-2.53 02	-3.04 02
13	-3.10 02	-2.42 02		-3.05 02	-3.66 02
14	-3.67 02	-2.89 02		-3.63 02	-4.36 02
15	-4.30 02	-3.41 02		-4.27 02	-5.12 02
16	-4.99 02	-3.97 02		-4.96 02	-5.97 02
17	-5.75 02	-4.59 02		-5.71 02	-6.90 02
18	-6.57 02	-5.27 02		-6.53 02	-7.92 02
19	-7.68 02	-5.99 02		-7.40 02	-8.94 02
20	-8.57 02	-6.77 02	-7.93 02	-8.35 02	-1.00 03
21	-9.58 02	-7.60 02		-9.35 02	-1.13 03
22	-1.06 03	-8.48 02		-1.04 03	-1.24 03
23	-1.18 03	-9.43 02		-1.16 03	-1.36 03
24	-1.30 03	-1.04 03		-1.28 03	-1.48 03
25	-1.43 03	-1.15 03		-1.40 03	-1.60 03
26	-1.56 03	-1.26 03		-1.54 03	-1.73 03
27	-1.71 03	-1.38 03		-1.68 03	-1.87 03
28	-1.86 03	-1.51 03		-1.83 03	-2.01 03
29	-2.01 03	-1.64 03		-1.99 03	-2.16 03
30	-2.18 03	-1.78 03		-2.15 03	-2.31 03

* -9.22 00 = -9.22×10^{0}, -1.77 01 = - -1.77×10^{1}, etc.

References

1. Abramowitz, M. and Stegun, I.A., Eds.: Handbook of Mathematical Functions. New York, Dover 1965.

2. Aronszajn, N.: Approximation Methods for Eigenvalues of Completely Continuous Symmetric Operators. Proc. Symp. Spectral Theory and Differential Problems. Oklahoma A & M College, Stillwater, OK, 1951, pp. 179-202.

3. Balslev, E.: Spectral Theory of Schrödinger Operators of Many-Body systems with Permutation and Rotation Symmetries. Ann. Phys. 73(1972), 49-107.

4. Bazley, N.W.: Lower Bounds for Eigenvalues with Application to the Helium Atom. Proc. Nat. Acad. Sci. U.S.A. 45(1959), 850-853.

5. Beattie, C.A.: The Computation of Convergent Lower Bounds in Quantum Mechanical Eigenvalue Problems. These Proceedings.

6. Condon, E.U. and Shortley, G.H.: The Theory of Atomic Spectra. Cambridge, University Press 1935.

7. Desclaux, J.P.: Relativistic Dirac-Fock Expectation Values with Z = 1 to Z = 120. Atomic Data and Nuclear Data Tables 12(1973), 311-406.

8. Flügge, S.: Practical Quantum Mechanics, Vol. 1. Berlin, Springer 1971.

9. Flügge, S.: Practical Quantum Mechanics, Vol. 2. Berlin, Springer 1971.

10. Fox, D. W.: Lower Bounds for Eigenvalues with Displacement of Essential Spectra. SIAM J. Math. Anal. 3(1972), 617-624.

11. Fox, D.W. and Sigillito, V.G.: Bounds for Energies of Radial Lithium. Z. Angew. Math. Phys. 23(1972), 392-411.

12. Fox, D.W. and Sigillito, V.G.: Lower and Upper Bounds to Energies of Radial Lithium. Chem. Phys. Lett. 13 (1972), 85-87.

13. Freund, D.E.: Upper and Lower Bounds to Two and Three-Electron Systems. Univ. of Delaware, Ph.D. Thesis, Dept. of Physics, June 1982.

14. Froese-Fischer, C.: The Hartree-Fock Method for Atoms. New York, Wiley 1977.

15. Grosse, H., Hertel, P., and Thirring, W.: Lower Bounds to the Energy Levels of Atomic and Molecular Systems. Acta Phys. Austriaca 49(1978), 89-112.

16. Hertel, P., Lieb, E.H., and Thirring, W.: Lower Bound to the Energy of Complex Atoms. J. Chem. Phys. 62(1975), 3355-3356.

17. Hill, R.N.: Proof that the H^- Ion has Only One Bound State . Phys. Rev. Lett. 38(1977), 643-646.

18. Kato, T.: Perturbation Theory for Linear Operators. Berlin, Springer 1966.

19. Newton, R.G.: Scattering Theory for Waves and Particles. New York, McGraw-Hill 1966.

20. Reed, M. and Simon, B.: Methods of Modern Mathematical Physics, II: Fourier Analysis, Self-Adjointness. New York, Academic Press 1975.

21. Russell, D.M.: The Calculation of Lower Bounds to Atomic Energies. Univ. of Arizona, Ph.D. Thesis, Program in Applied Mathematics, July 1983.

22. Simon, B.: On the Infinitude or Finiteness of the Number of Bound States of an N-Body Quantum System, I. Helv. Phys. Acata 43(1970), 607-630.

23. Thirring, W.: A Course in Mathematical Physics, Vol. 3. Berlin, Springer 1979.

24. Uchiyama, J.: Finiteness of the Number of Discrete Eigenvalues of the Schrodinger Operator for a Three Particle System. Publ. Res. Inst. Math. Sci. A5 (1969), 51-63.

25. Zhislin, G.M.: Discussion of the Spectrum of the Schrödinger Operator for Systems of Many Particles. Trudy Moskov. Mat Obšč. 9(1960), 81-128.

26. Zhislin, G.M.: On the Finiteness of the Discrete Spectrum of the Energy Operator of Negative Atomic and Molecular Ions. Teoret. Mat. Fiz. 21(1971), 332-341; Theoret. and Math. Phys. 7(1971), 571-578.

Department of Mathematics and Program in Applied Mathematics
University of Arizona
Tucson, Arizona 85721

(Current address of D. M. Russell: Department of Mathematical Sciences, University of North Carolina at Wilmington, Wilmington, North Carolina 28403)

ISNM, Vol. 69
Numerical Treatment of
Eigenvalue Problems, Vol. 3

UEBER EIGENWERTE SYMMETRISCHER MEMBRANEN

Joseph Hersch

Some simple examples illustrate the fact that useful a priori bounds for membrane eigenvalues can be obtained from isoperimetric inequalities and simple monotonicity arguments. In the present case of symmetric domains, it is important to know their conformal radius (mapping radius) at the center of symmetry. For some symmetric domains there is an exact elementary ratio of the conformal radii. Also, some symmetric membranes have the same first eigenvalue.

1. Die Prinzipien von Rayleigh und Poincaré, ein Monotoniesatz

1.1. Wir betrachten die Eigenwerte $\lambda_1 < \lambda_2 \leq \lambda_3 \leq \dots$ einer homogenen Membran auf einem ebenen Gebiet G, welche längs der Randkurve ∂G eingespannt ist:

$$\Delta u + \lambda u = 0 \quad \text{in} \quad G \;, \qquad u = 0 \quad \text{längs} \quad \partial G \;.$$

1.2. Diese Eigenwerte sind die stationären Werte des Rayleigh-Quotienten $R[v] := D(v)/\iint_G v^2 dA$, wo $dA = dxdy$ das Flächenelement und $D(v)$ das Dirichlet-Integral $\iint_G |\text{grad } v|^2 dA$ bezeichnet. Dabei werden nur solche Funktionen v zugelassen, die längs ∂G verschwinden.

1.3. Das *Rayleigh-Prinzip* [7] besagt: $\lambda_1 = \mathrm{Min}_v R[v]$.

Das *Poincaré-Prinzip* [6] charakterisiert die höheren Eigenwerte durch

$$\lambda_n = \mathrm{Min}_{L_n} \max_{v \in L_n} R[v] ,$$

L_n bezeichnet hier irgendeinen n-dimensionalen linearen Raum von zulässigen Funktionen.

1.4. Aus diesen Charakterisierungen der Eigenwerte als Minima folgt insbesondere der folgende *Monotoniesatz: Bei Einschränkung des Raumes der zulässigen Funktionen werden alle Eigenwerte grösser.* (D.h. genauer: Kein Eigenwert wird kleiner.)

2. Die isoperimetrischen unteren Schranken von Rayleigh-Faber-Krahn und von M.-Th. Kohler

2.1. Der klassischste isoperimetrische Satz über Eigenwerte einer Membran ist derjenige von Rayleigh-Faber-Krahn [7]: *Unter allen Membranen mit gegebenem Flächeninhalt* A *hat die kreisförmige den tiefsten Grundton,* das heisst $\lambda_1 \geqslant \pi j_o^2/A$ ($j_o = 2.4048...$) .

2.2. Die *Torsionssteifigkeit* P eines *einfach zusammenhängenden* ebenen Gebietes G kann so definiert werden: Sei t die Lösungsfunktion des Randwertproblems

$$\Delta t = -1 \text{ in } G , \quad t = 0 \text{ längs } \partial G ,$$

dann ist $P = 4D(t) = 4\iint_G t\, dA$.

Es gilt der isoperimetrische Satz von Saint-Venant und Pólya [7]: *Unter allen einfach zusammenhängenden Gebieten mit gegebenem Flächeninhalt* A *hat die Kreisscheibe die grösste Torsionssteifigkeit,* das heisst $P \leqslant A^2/2\pi$.

2.3. Folgender isoperimetrischer Satz wurde 1951 von Pólya und Szegö [7] vermutet und (zum ersten Mal 1975) von M.-Th. Kohler [3,4] auf zwei verschiedenen Wegen bewiesen: Die reine Zahl $P\lambda_1^2$ *ist bei der Kreisscheibe am kleinsten,* das heisst $P\lambda_1^2 \geqslant \pi j_o^4/2$.

Wie der oben erwähnte Satz von Saint-Venant und Pólya zeigt, ist dies eine Verschärfung des Satzes von Rayleigh-Faber-Krahn. Numerisch erhält man viel schärfere untere Schranken für λ_1 .

3. Isoperimetrische obere Schranken mit Hilfe des Abbildungsradius

3.1. Sei G ein *einfach zusammenhängendes* Gebiet in der z-Ebene und sei O irgend ein (innerer) Punkt von G . Der *Abbildungsradius* (oder konforme Radius) $R_o(G)$ *des Gebietes* G *im Punkt* O ist nach Definition der Radius derjenigen Kreisscheibe $|w| < R$, auf die G konform und schlicht durch eine *in* O *normierte* Abbildung $w(z)$ abgebildet werden kann: $w(O) = 0$, $w'(O) = 1$.

Der "maximale Abbildungsradius" von G ist $\dot{R}(G) = \max_{O\in G} R_o(G)$.

<u>3.2.</u> Ein isoperimetrischer Satz von Pólya und Szegö [7] besagt: Die reine Zahl $\lambda_1 \dot{R}^2$ *ist bei der Kreisscheibe am grössten*, das heisst $\lambda_1 \dot{R}^2 \leqslant j_o^2$.

<u>3.3.</u> Ein isoperimetrischer Satz von Pólya und Schiffer [6] besagt: *Für jede natürliche Zahl* N *ist* $(\frac{1}{\lambda_1} + \frac{1}{\lambda_2} + \ldots + \frac{1}{\lambda_N}) \frac{1}{\dot{R}^2}$ *bei der Kreisscheibe am kleinsten.*

<u>3.4.</u> Ein Gebiet G heisst "*symmetrisch von der Ordnung* q " um den Punkt O , wenn G bei einer Drehung um O um den Winkel $\frac{2\pi}{q}$ unverändert bleibt. Nach einem Satz von Pólya [5] gilt: *Je höher die Symmetrie* q *des einfach zusammenhängenden Gebietes* G , *desto grösser die Zahl* p(q) *der Eigenwerte* $\lambda_1, \lambda_2, \ldots, \lambda_{p(q)}$, *für die* $\lambda_k R_o^2(G)$ *höchstens so gross ist wie für eine Kreisscheibe mit Zentrum* O . Pólya erhält p(3) = 3 , p(5) = 6 , p(7) = 10 , p(9) = 15 , p(11) = 19 . Einige Ergänzungen sind in [1] aufgeführt: p(2) = 2 , p(4) = 4 .

4. Naheliegende Vermutungen für reguläre n-Ecke

<u>4.1.</u> Abschnitt 2.1 legt die Vermutung nahe, dass $\lambda_1 A$ eine monoton *abnehmende* Funktion von n sei.

<u>4.2.</u> Abschnitt 2.3 legt die Vermutung nahe, dass $P\lambda_1^2$ eine monoton *abnehmende* Funktion von n sei.

<u>4.3.</u> Abschnitt 3.2 legt die Vermutung nahe, dass $\lambda_1 R_o^2$ eine monoton *zunehmende* Funktion von n sei.

4.4. Abschnitt 3.3 legt die Vermutung nahe, dass für jede feste natürliche Zahl N der Ausdruck $(\frac{1}{\lambda_1} + \frac{1}{\lambda_2} + \ldots + \frac{1}{\lambda_N}) \frac{1}{R_o^2}$ eine monoton *abnehmende* Funktion von n sei.

5. Beispiele

5.1. Grundton eines regulären Sechsecks mit Seite 1

Der Satz von Rayleigh-Faber-Krahn (Abschnitt 2.1) gibt $\lambda_1 > 6.9930$.

Die Verschärfung von M.-Th. Kohler (Abschnitt 2.3) gibt $\lambda_1 > 7.1214$. Dafür verwenden wir den Wert $P = 1.0359$ ([7], Seite 258).

Der Satz von Pólya-Szegö (Abschnitt 3.2) gibt $\lambda_1 < 7.1636$. Dafür verwenden wir den Wert $R_o = 0.89850$ ([7], Seite 258). Also $7.1214 < \lambda_1 < 7.1636$.

Die *Vermutungen vom Abschnitt 4*, angewendet auf den *Vergleich mit dem Quadrat*, ergeben folgendes:

Vermutung 4.1: $\lambda_1 \overset{?}{<} 7.5976$ stimmt;

Vermutung 4.2: $\lambda_1 \overset{?}{<} 7.2722$ stimmt;

Vermutung 4.3: $\lambda_1 \overset{?}{>} 7.1127$ stimmt.

5.2. Zweiter = dritter Eigenwert desselben regulären Sechsecks

Wegen $q = 6 \geqslant 3$ haben wir nach Pólya (siehe hier Abschnitt 3.4) $\lambda_2 = \lambda_3 < 18.187$.

Dieses λ_2 ist auch der erste Eigenwert λ_1^T des halben regulären Sechsecks T (siehe Figur). Eine untere Schranke für ihn erhalten wir durch eine einfache Monotonie-Ueberlegung (Abschnitt 1.4). Dies ist nämlich auch der zweite Eigenwert

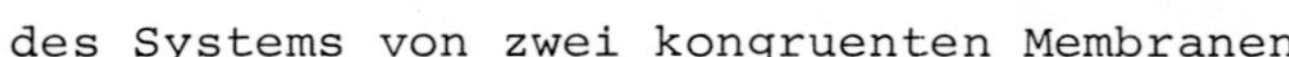

des Systems von zwei kongruenten Membranen

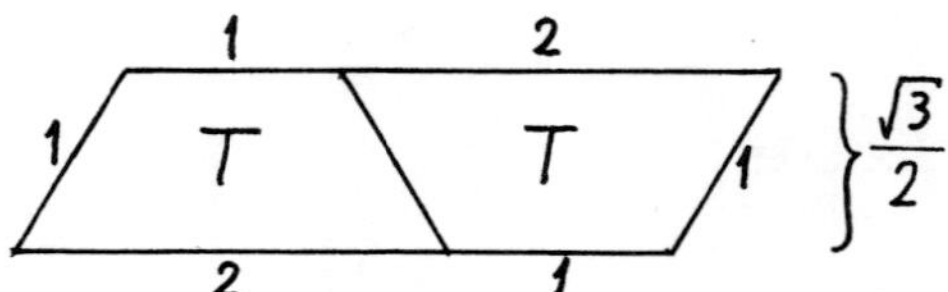

Nach Abschnitt 1.4 ist er also grösser als der *zweite* Eigenwert der durch *Identifikation* der linken und rechten schrägen Seiten definierten zweifach zusammenhängenden Membran Z .

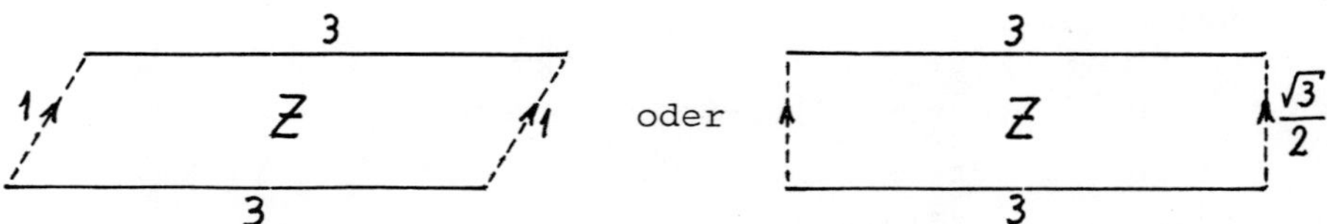

Die untere Schranke, die wir bekommen, ist also der erste Eigenwert des Rechtecks mit Seiten 3/2 und $\sqrt{3}/2$, das heisst $16\pi^2/9 \simeq 17.546$. Wir haben also $17.546 < \lambda_1^T < 18.187$.

5.3. Vierter = fünfter Eigenwert desselben regulären Sechsecks

Wegen $q = 6 \geq 5$ haben wir nach Pólya (siehe hier Abschnitt 3.4) $\lambda_4 = \lambda_5 < 32.670$.

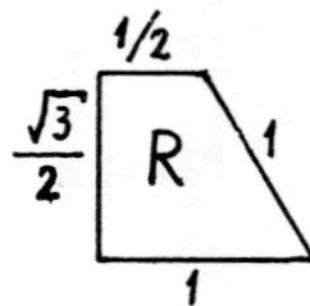

Dieses λ_4 ist auch der erste Eigenwert λ_1^R des gezeichneten rechteckigen Trapezes R . Er ist auch der zweite Eigenwert des Systems von zwei kongruenten Membranen

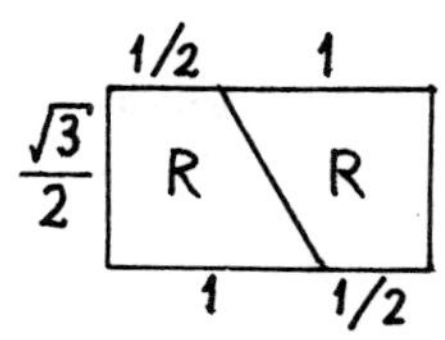

also nach dem Monotoniesatz (Abschnitt 1.4) höher als der *zweite* Eigenwert der rechteckigen eingespannten Membran mit Seiten 3/2 und $\sqrt{3}/2$, d.h. der erste Eigenwert des Rechtecks mit Seiten 3/4 und $\sqrt{3}/2$, also $28\pi^2/9 \simeq 30.705$. Wir haben also

$30.705 < \lambda_1^R < 32.670$.

5.4. *Die Vermutung 4.4*, angewendet auf den *Vergleich mit dem Quadrat*, ergibt für die Eigenwerte des Sechsecks folgende Ungleichungen:

(a) $\frac{1}{\lambda_1} + \frac{1}{\lambda_2} \overset{?}{<} 0.19683$;

nach 5.1 und 5.2 haben wir nur

$$0.19458 < \frac{1}{\lambda_1} + \frac{1}{\lambda_2} < 0.19741 .$$

(b) $\frac{1}{\lambda_1} + \frac{1}{\lambda_2} + \frac{1}{\lambda_3} \overset{?}{<} 0.25307$;

nach 5.1 und 5.2 haben wir nur

$$0.24956 < \frac{1}{\lambda_1} + \frac{1}{\lambda_2} + \frac{1}{\lambda_3} < 0.25441 .$$

(c) $\frac{1}{\lambda_1} + \frac{1}{\lambda_2} + \frac{1}{\lambda_3} + \frac{1}{\lambda_4} \overset{?}{<} 0.28822$; *dies stimmt*,

nach 5.1, 5.2 und 5.3 haben wir nämlich

$$0.28017 < \frac{1}{\lambda_1} + \frac{1}{\lambda_2} + \frac{1}{\lambda_3} + \frac{1}{\lambda_4} < 0.28698 .$$

(d) $\frac{1}{\lambda_1} + \frac{1}{\lambda_2} + \frac{1}{\lambda_3} + \frac{1}{\lambda_4} + \frac{1}{\lambda_5} \overset{?}{<} 0.32336$; *dies stimmt*

wieder, nach 5.1, 5.2 und 5.3 haben wir nämlich

$$0.31078 < \frac{1}{\lambda_1} + \frac{1}{\lambda_2} + \frac{1}{\lambda_3} + \frac{1}{\lambda_4} + \frac{1}{\lambda_5} < 0.31954 .$$

5.5. Grundton des Sterns von David

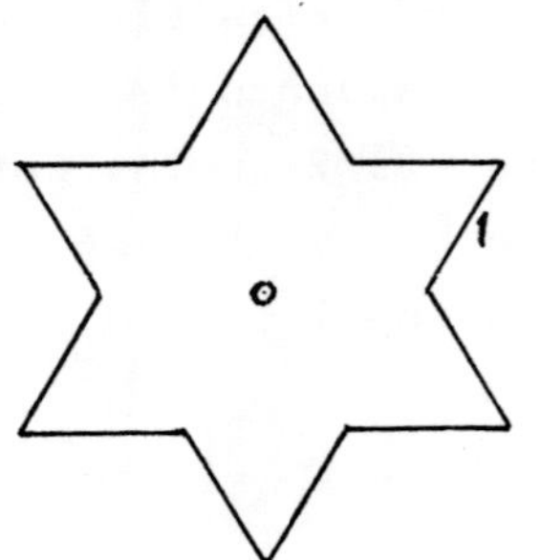

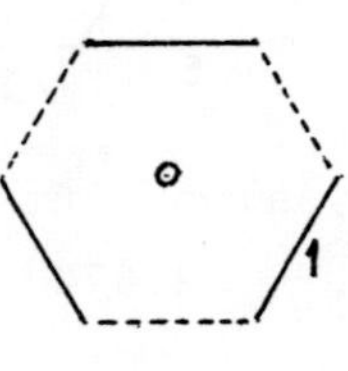

Der erste Eigenwert λ_1 einer eingespannten Membran auf einem David-Stern ist [2] *genau* gleich demjenigen einer Membran auf einem regulären Sechseck mit derselben Seitenlänge, welche aber *längs der Randstrecken abwechselnd eingespannt* $(u = 0)$ *und frei* $(\partial n/\partial n = 0)$ *ist.*

Wie anderswo gezeigt werden wird, ist *das Verhältnis* des Abbildungsradius des David-Sterns in seinem Zentrum zu demjenigen des regulären Sechsecks mit derselben Seitenlänge *genau* gleich $2^{1/3}$. (Spezialfall von allgemeineren Beziehungen zwischen Abbildungsradien.) Da der Abbildungsradius R_o eines regulären Sechsecks mit Seitenlänge 1 tabelliert ist ([7], Seite 258) und 0.89850 beträgt, finden wir nach Pólya und Szegö (Abschnitt 3.2 hier) $\lambda_1 < 4.5128$.

Der Satz von Rayleigh-Faber-Krahn (Abschnitt 2.1) gibt nur die schwache untere Schranke $\lambda_1 > 3.4965$. Zur Konstruktion der schärferen unteren Schranke nach M.-Th. Kohler (Abschnitt 2.3 hier) sollte man zunächst die Torsionssteifigkeit P des David-Sterns berechnen.

Der Satz von Pólya (Abschnitt 3.4 hier) gibt auch obere Schranken für $\lambda_2, \lambda_3, \ldots, \lambda_6$.

Literatur

[1] Hersch, J.: On symmetric membranes and conformal radius: some complements to Pólya's and Szegö's inequalities. Arch. Rat. Mech. Anal. 20 (1965), 378-390.

[2] Hersch, J.: Membranes symétriques d'égale fréquence fondamentale. Z. Angew. Math. Phys. 30 (1979), 220-233.

[3] Kohler-Jobin, M.-Th.: Une méthode de comparaison isopérimétrique de fonctionnelles de domaines de la physique mathématique. Z. Angew. Math. Phys. 29 (1978), 757-776.

[4] Kohler-Jobin, M.-Th.: Isoperimetric monotonicity and isoperimetric inequalities of Payne-Rayner type for the first eigenfunction of the Helmholtz problem. Z. Angew. Math. Phys. 32 (1981), 625-646.

[5] Pólya, G.: On the characteristic frequencies of a symmetric membrane. Math. Z. 63 (1955), 331-337.

[6] Pólya, G. and Schiffer, M.: Convexity of functionals by transplantation. J. d'Analyse Math. 3 (2ème partie) (1953/54), 245-345.

[7] Pólya, G. and Szegö, G.: Isoperimetric inequalities in mathematical physics, Princeton Univ. Press 1951.

Joseph Hersch

Mathematik

ETH-Zentrum

CH - 8092 Zürich/Schweiz

ISNM, Vol. 69
Numerical Treatment of
Eigenvalue Problems, Vol. 3

ZUR ANWENDUNG DER THEORIE POSITIVER OPERATOREN AUF EIGENWERTAUFGABEN MIT GEWÖHNLICHEN DIFFERENTIALGLEICHUNGEN

Peter Paul Klein

To a class of general eigenvalue problems of the form $My = \lambda Ny$ with linear ordinary differential operators M and N of orders 4 and 2 respectively the reduction method of J. Schröder [9] is applied in order to prove that $M^{-1}N$ is a positive operator. By means of the theory of positive operators the existence of a nonnegative eigenfunction belonging to a positive eigenvalue, minimal in modulus, of the original eigenvalue problem is concluded. The quotient theorem for positive operators is used for the enclosure of this eigenvalue and compared with another enclosure theorem using quotients given by W. Held [5].

<u>Bezeichnungen:</u> Die Abkürzungen RWA, RB, wes. RB, EWA, EW, EF stehen für Randwertaufgabe, Randbedingung, wesentliche Randbedingung, Eigenwertaufgabe, Eigenwert, Eigenfunktion. $C_k[x_1,x_2]$ sei die Menge der im Intervall $[x_1,x_2]$ k-mal stetig differenzierbaren Funktionen.

Aus der Theorie halbgeordneter linearer Räume werden folgende Bezeichnungen und Begriffe verwendet (vgl. z. B. [9]). $(R,\leq)$ sei ein halbgeordneter linearer Raum. Θ bezeichne das Nullelement eines linearen Raumes. Für $u \in R$ sei gesetzt:

$$u > \Theta \iff u \geq \Theta,\ u \neq \Theta.$$

Ferner sei eine strenge Halbordnung $\succ$ eingeführt durch: $u \succ \Theta \iff$ für alle $v \in R$ existiert ein $n \in \mathbb{N}$ mit $nu + v \geq \Theta$. $(S,\leq)$ sei ein weiterer halbgeordneter linearer Raum. Ein Operator $A : R \to S$

heißt positiv, wenn für alle $u \in R$ gilt: $u \geq \Theta \Rightarrow Au \geq \Theta$. A heißt punktweise streng positiv, wenn für alle $u \in R$ gilt: $u > \Theta \Rightarrow Au \succ \Theta$. Ein linearer Operator $M : R \to S$ heißt inverspositiv oder von monotoner Art, wenn für alle $u \in R$ gilt: $Mu \geq \Theta \Rightarrow u \geq \Theta$.

Problemstellung

Gegeben sei die allgemeine EWA

$$My = \lambda Ny, \quad U_j[y] = 0, \quad V_j[y] = 0 \quad (j=1,2)$$

mit den Differentialausdrücken

$$My = (p_2y'')'' - (p_1y')' + p_oy, \quad p_j \in C_j[x_1,x_2] \quad (j=0,1,2)$$
$$Ny = \quad - (q_1y')' + q_oy, \quad q_j \in C_j[x_1,x_2] \quad (j=0,1)$$

und den Randausdrücken U_j, V_j gemäß Tabelle 1; dabei sind die hinter "v" stehenden Randausdrücke nur zugelassen, wenn die spezielle EWA vorliegt, d. h. wenn $q_1(x) \equiv 0$ gilt.

Fall	$U_1[y]$	$U_2[y]$	$V_1[y]$	$V_2[y]$
1.1	$y(x_1)$	$B_1y(x_1) \vee y'(x_1)$	$y(x_2)$	$B_1y(x_2) \vee y'(x_2)$
1.2.1	$y(x_1)$	$B_1y(x_1) \vee y'(x_1)$	$y'(x_2) \vee B_1y(x_2)$	$B_oy(x_2)$
1.2.2	$y'(x_1) \vee B_1y(x_1)$	$B_oy(x_1)$	$y(x_2)$	$B_1y(x_2) \vee y'(x_2)$
1.3	$y'(x_1) \vee B_1y(x_1)$	$B_oy(x_1)$	$y'(x_2) \vee B_1y(x_2)$	$B_oy(x_2)$

$$\text{mit } B_1y(x_i) = g_{1i}y'(x_i) + (-1)^i p_2y''(x_i)$$
$$B_oy(x_i) = g_{oi}y(x_i) + (-1)^i((-p_2y'')'(x_i) + p_1y'(x_i)) \quad (i=1,2)$$

Tabelle 1: Randausdrücke

Es wird definiert:

$$R = \{u \in C_4[x_1,x_2] \mid U_j[u] = 0, V_j[u] = 0 \ (j=1,2)\}, \quad S = C_o[x_1,x_2]$$

$$\tilde{R} = \begin{cases} S & \text{falls } q_1(x) \equiv 0 \\ \{u \in C_2[x_1,x_2] \mid u \text{ erfüllt die wes. RB von } R\} & \text{sonst.} \end{cases}$$

In den wes. RB von R tritt keine Ableitungsordnung >1 auf. Hat die homogene RWA $My = \Theta$, $y \in R$ nur die triviale Lösung, so existiert die zugehörige Greensche Funktion G und die EWA

$$My = \lambda Ny, \quad y \in R \quad (1)$$

ist äquivalent zu der EWA

$$y(x) = \lambda \int_{x_1}^{x_2} G(x,\xi)Ny(\xi)d\xi, \quad y \in \tilde{R}. \tag{2}$$

Da die RWA $My = \Theta$, $y \in R$ selbstadjungiert ist, gilt $G(x,\xi) = G(\xi,x)$ in $x_1 \leq x,\xi \leq x_2$. Aus der EWA (2) folgt im Fall $q_1(x) \not\equiv 0$ durch partielle Integration die EWA

$$\begin{aligned} &y = \lambda Ay, \ y \in S \text{ mit } Ay(x) = \int_{x_1}^{x_2} N_\xi G(x,\xi)y(\xi)d\xi, \\ &N_\xi G(x,\xi) = -(q_1(\xi)G_\xi(x,\xi))_\xi + q_o(\xi)G(x,\xi). \end{aligned} \tag{3}$$

Die bei der partiellen Integration auftretenden Randterme verschwinden, da $y \in \tilde{R}$ und G als Funktion von ξ die wes. RB von R erfüllen. Umgekehrt folgt aus der EWA (3) auch die EWA (2), da für A gilt $A : S \to \tilde{R}$ und deshalb die partielle Integration rückgängig gemacht werden kann. Damit sind die EWAn (1) und (3) äquivalent, sofern die RWA $My = \Theta$, $y \in R$ nur die triviale Lösung hat.

Im folgenden sollen hinreichende Bedingungen dafür angegeben werden, daß die Kernfunktion N_ξ nichtnegativ ist. Bei dem nachstehenden Beispiel aus [3] läßt sich dies unmittelbar erkennen:

$$(\alpha y'')'' = \lambda\rho(-Iy''+Fy) \text{ in } [0,l], \ y(0) = y''(0) = y(l) = y''(l) = 0$$
$$\text{mit } \alpha(x) > 0 \text{ in } 0 \leq x \leq l \text{ und } \rho,F,I \in \mathbb{R}.$$

Für die zu $M : R \to S$ gehörige Greensche Funktion (vgl. [3], S. 453) gilt: $G_{\xi\xi}(x,\xi) = -\frac{x(l-\xi)}{l\cdot\alpha(\xi)}$ für $x \leq \xi$,

$G_{\xi\xi}(x,\xi) = G_{\xi\xi}(\xi,x)$ für $\xi \leq x$. Wegen $G(x,0) = G(x,l) = 0$ und $-G_{\xi\xi}(x,\xi) \geq 0$ in $0 \leq x,\xi \leq l$ folgt dort $G(x,\xi) \geq 0$. Damit gilt für nichtnegative Konstanten ρ,F,I $N_\xi G(x,\xi) = \rho[-IG_{\xi\xi}(x,\xi) + FG(x,\xi)] \geq 0$ in $0 \leq x,\xi \leq l$.

§ 1 Das Reduktionsverfahren

Zur Behandlung des allgemeinen Falls soll das Reduktionsverfahren von J. Schröder (vgl. [9], S. 95 ff) angewendet werden. Dazu wird ein weiterer Raum eingeführt: $Y = \{u \in C_2[x_1,x_2] \mid u \text{ erfüllt die wes. RB von } R\}$. Die Räume R,S,Y seien mit der natürlichen

Halbordnung $\leq$ versehen. Die strenge Halbordnung in S sei mit $\succ$ bezeichnet. Es gilt: $U \in S$, $U \succ \Theta \iff U(x) > 0$ in $x_1 \leq x \leq x_2$. Zur Charakterisierung der strengen Halbordnung in R sei das Funktional l_x für $x \in [x_1,x_2]$ eingeführt (vgl. [9], S. 145): Es sei $n_i \in \mathbb{N}_o$ die kleinste Ableitungsordnung, so daß für ein $u_i \in R$ gilt $u_i^{(n_i)}(x_i) \neq 0$ $(i=1,2)$. Sei $u \in R$: $u \succ \Theta \iff l_x u > 0$ für $x \in [x_1,x_2]$ mit $l_{x_1} u = u^{(n_1)}(x_1)$, $l_x u = u(x)$ für $x_1 < x < x_2$,

$l_{x_2} u = (-1)^{n_2} u^{(n_2)}(x_2)$. Dieselbe Charakterisierung gilt für Y, da in Y die wes. RB von R gelten.

<u>VORAUSSETZUNG (V1):</u> Für die Funktionen $\psi, \phi \in C_4[x_1,x_2]$ gelte: ψ erfüllt die wes. RB von R bei x_1, ϕ bei x_2;

$$l_x\psi > 0 \text{ für } x \in [x_1,x_2), \quad l_x\phi > 0 \text{ für } x \in (x_1,x_2].$$

<u>DEFINITION 1.1:</u> Sei $U \in S$,

$$PU(x) = \int_{x_1}^{x_2} \Gamma(x,t)U(t)dt \text{ mit } \Gamma(x,t) = \begin{cases} \psi(x)\phi(t) & x \leq t \\ \psi(t)\phi(x) & t \leq x. \end{cases}$$

Ist (V1) erfüllt, so ist P eine Abbildung von S in Y mit der Eigenschaft: $U \in S$, $U > \Theta \Rightarrow PU \succ \Theta$, d. h. P ist ein punktweise streng positiver Operator.

<u>SATZ 1.2:</u> Die Operatoren $M : R \to S$ und $PM : R \to Y$ seien invertierbar. Dann gilt $M^{-1}N = (PM)^{-1}PN$. Ist zusätzlich $PN : R \to Y$ ein positiver und PM ein inverspositiver Operator, so ist $M^{-1}N$ ein positiver Operator von R in sich.

<u>Beweis:</u> Die Abbildung $M : R \to S$ ist surjektiv. Da M^{-1} existiert, läßt sich mit Hilfe der zugehörigen Greenschen Funktion zu jedem $r \in S$ eine Lösung der RWA $Mu = r$, $u \in R$ angeben. Wegen $M(R) = S$ folgt $PM(R) = P(S)$. Aus $P = P(MM^{-1}) = (PM)M^{-1}$ und der Tatsache, daß $(PM)^{-1}$ auf $P(S) = PM(R)$ existiert, ergibt sich: $(PM)^{-1}P = M^{-1}$ also $(PM)^{-1}PN = M^{-1}N$.

Mit dem in [9] angegebenen Monotoniesatz für lineare Operatoren kann die Inverspositivität von PM und in Verbindung damit auch die von M nachgewiesen werden. Der Operator $PM : R \to Y$ ist inverspositiv, wenn gilt:

(I) Für alle $u \in R$, $u \geq \Theta$, $PMu \succ \Theta \Rightarrow u \succ \Theta$

(II) Es gibt ein $z \in R$, $z \geq \Theta$ mit $PMz \succ \Theta$.

Der Operator $M : R \to S$ ist inverspositiv, wenn gilt:

(I') Für alle $u \in R$, $u \geq \Theta$, $Mu > \Theta \Rightarrow u \succ \Theta$

(II') Es gibt ein $z \in R$, $z \geq \Theta$ mit $Mz > \Theta$.

Besitzt PM die Eigenschaft (I), so heißt PM schwach präinverspositiv; besitzt M die Eigenschaft (I'), so heißt M stark präinverspositiv. Ein Element z, welches die Eigenschaft (II) hat, heißt stark majorisierendes Element für PM; ein Element z, welches Eigenschaft (II') hat, heißt schwach majorisierendes Element für M. Ist (V1) erfüllt, so besteht zwischen (I) und (I') sowie (II) und (II') folgender Zusammenhang: Aus (I) folgt (I') und aus (II') folgt (II). Damit ergibt sich:

<u>SATZ 1.3:</u> Es gelte (V1) sowie (I) und (II'). Dann sind die Operatoren $M : R \to S$ und $PM : R \to Y$ inverspositiv.

Durch partielle Integration ergibt sich für $u \in C_4[x_1,x_2]$:

$$PMu(x) = -p(x)p_2(x)u''(x) + p'(x)p_2(x)u'(x) + g(x)u(x) - \tag{4}$$
$$- \phi(x)[b_1(\psi,u) - \int_{x_1}^{x} u(t)M\psi(t)dt] - \psi(x)[b_2(\phi,u) - \int_{x}^{x_2} u(t)M\phi(t)dt]$$

mit $p = W(\phi,\psi) = \phi\psi' - \phi'\psi$

$g = \psi[(p_2\phi'')' - p_1\phi'] - \phi[(p_2\psi'')' - p_1\psi']$

$$b_1(\psi,u) = \psi(x_1)B_o u(x_1) + \psi'(x_1)B_1 u(x_1) - u'(x_1)B_1\psi(x_1) - u(x_1)B_o\psi(x_1)$$

$$b_2(\phi,u) = \phi(x_2)B_o u(x_2) + \phi'(x_2)B_1 u(x_2) - u'(x_2)B_1\phi(x_2) - u(x_2)B_o\phi(x_2).$$

Dabei bezeichnet $W(\phi,\psi)$ die Wronskische Determinante von ϕ und ψ; $b_i(\cdot,\cdot)$ bezeichnet eine Randbilinearform an der Stelle x_i $(i=1,2)$ mit den Randausdrücken B_o, B_1 gemäß Tabelle 1. Ferner gilt für $q_1(x) \neq 0$:

$PNu(x) = p(x)q_1(x)u(x) + Qu(x)$ mit

$$Qu(x) = \phi(x)[\psi(x_1)q_1u'(x_1) - \psi'(x_1)q_1u(x_1) + \int_{x_1}^{x} u(t)N\psi(t)dt] +$$

$$+ \psi(x)[-\phi(x_2)q_1u'(x_2) + \phi'(x_2)q_1u(x_2) + \int_x^{x_2} u(t)N\phi(t)dt].$$

VORAUSSETZUNG (V2): $p(x) \geq 0$, $q_1(x) \geq 0$ in $x_1 \leq x \leq x_2$; $M\psi \leq \Theta$, $N\psi \geq \Theta$; $M\phi \leq \Theta$, $N\phi \geq \Theta$.
Ferner seien gemäß Tabelle 2 die Ungleichungen für die mit ψ und ϕ gebildeten Randausdrücke B_o, B_1 erfüllt.

BEMERKUNG 1.4: Die RB $U_2[y] = y'(x_1) = 0$, $V_2[y] = y'(x_2) = 0$ wurden nur im Fall $q_1(x) \equiv 0$ zugelassen, da bei einer allgemeinen EWA mit zwei wes. RB am selben Intervallende (V1) und (V2) nicht gleichzeitig erfüllbar zu sein brauchen, wie folgendes Beispiel aus [3] zeigt:
$y^{IV} = \lambda[-(x^2y')' + 2y]$ in $[0,1]$, $y'(0) = y'''(0) = y(1) = y'(1) = 0$.
Eine Funktion $\phi \in C_4[0,1]$ mit $N\phi \geq \Theta$, $\phi(1) = \phi'(1) = 0$ und $\phi > \Theta$ läßt sich hier nicht angeben.

SATZ 1.5: Unter den Voraussetzungen (V1), (V2) gilt:
1. $PN : R \to Y$ ist ein positiver Operator und $b_1(\psi,\cdot)$, $b_2(\phi,\cdot)$ sind positive Funktionale auf R.
2. Der Operator $PM : R \to Y$ ist schwach präinverspositiv, d. h. für alle $u \in R$ gilt: $u \geq \Theta$, $PMu \succ \Theta \Rightarrow u \succ \Theta$.

Beweis: 1. Sei $u \in R$, $u \geq \Theta$. Die Randterme in $Qu(x)$ verschwinden wegen $U_1[u] = V_1[u] = 0$, $U_1[\psi] = V_1[\phi] = 0$. Die Integralterme in $Qu(x)$ sind nichtnegativ wegen $u \geq \Theta$, $N\psi \geq \Theta$, $N\phi \geq \Theta$. Wegen $p(x) \geq 0$, $q_1(x) \geq 0$ in $x_1 \leq x \leq x_2$ folgt dann, daß $PN : R \to Y$ ein positiver Operator ist. Das Funktional $b_1(\psi,\cdot) : R \to \mathbb{R}$ ist positiv. Liegen in R zwei wes. RB bei x_1 vor, so gilt $b_1(\psi,u) = 0$; liegt eine bzw. keine wes. RB bei x_1 vor, so verschwinden in $b_1(\psi,u)$ alle Randterme bis auf einen bzw. zwei. Die übrigen Randterme sind auf Grund der in (V1) und (V2) gestellten Bedingungen nichtnegativ. Der Nachweis für $b_2(\phi,\cdot)$ verläuft entsprechend.
2. Für ein $u \in R$ gelte $u \geq \Theta$, $PMu \succ \Theta$, $u \not\succ \Theta$. Im Fall $u(\xi) = 0$ für ein $\xi \in (x_1,x_2)$ folgen zunächst als notwendige Bedingungen für ein inneres Minimum $u'(\xi) = 0$ und $u''(\xi) \geq 0$. Wegen (V2) ergibt sich dann aus Darstellung (4) $PMu(\xi) \leq 0$ im Widerspruch zu $PMu \succ \Theta$. Im Fall $u^{(n_1)}(x_1) = 0$ mit $0 \leq n_1 \leq 2$ folgt aus

Fall	Randverhalten von ψ und ϕ bei x_1		Randverhalten von ϕ und ψ bei x_2	
1.1	$\psi(x_1) = 0$	$B_1\psi(x_1) \leq 0 \vee \psi'(x_1) = 0$	$\phi(x_2) = 0$	$B_1\phi(x_2) \geq 0 \vee \phi'(x_2) = 0$
1.2.1	$\psi(x_1) = 0$	$B_1\psi(x_1) \leq 0 \vee \psi'(x_1) = 0$	$\phi'(x_2) = 0 \vee \begin{cases} B_1\phi(x_2) \geq 0 \\ B_1\psi(x_2) \geq 0 \end{cases}$	$B_o\phi(x_2) \leq 0$
1.2.2	$\psi'(x_1) = 0 \vee \begin{cases} B_1\psi(x_1) \leq 0 \\ B_1\phi(x_1) \leq 0 \end{cases}$	$B_o\psi(x_1) \leq 0$	$\phi(x_2) = 0$	$B_1\phi(x_2) \geq 0 \vee \phi'(x_2) = 0$
1.3	$\psi'(x_1) = 0 \vee \begin{cases} B_1\psi(x_1) \leq 0 \\ B_1\phi(x_1) \leq 0 \end{cases}$	$B_o\psi(x_1) \leq 0$	$\phi'(x_2) = 0 \vee \begin{cases} B_1\phi(x_2) \geq 0 \\ B_1\psi(x_2) \geq 0 \end{cases}$	$B_o\phi(x_2) \leq 0$

Tabelle 2: Randverhalten von ψ und ϕ

$$(PMu)^{(n_1)}(x_1) = \psi^{(n_1)}(x_1)\int_{x_1}^{x_2} \phi(t)Mu(t)dt =$$

nach partieller Integration

$$\psi^{(n_1)}(x_1)\{-b_2(\phi,u) - b_1(\phi,u) + \int_{x_1}^{x_2} u(t)M\phi(t)dt\}.$$

Hiermit ergibt sich wegen (V2) $(PMu)^{(n_1)}(x_1) \leq 0$ im Widerspruch zu $PMu \succ \Theta$. Im Fall $u^{(n_2)}(x_2) = 0$ mit $0 \leq n_2 \leq 2$ verläuft der Beweis entsprechend.

§ 2 Konstruktion eines majorisierenden Elementes für M

Es sei gesetzt: $\hat{M}u = (p_2u'')'' - (p_1u')'$ für $u \in R$.

VORAUSSETZUNG (V3): (a) Die EWA

$$\hat{M}u = \sigma u, \quad u \in R \tag{5}$$

besitze einen EW σ_o mit einer EF $u_o > \Theta$.
(b) Es gelte $p_o(x) > -\sigma_o$ in $x_1 \leq x \leq x_2$.

SATZ 2.1: Ist (V3) erfüllt, so gibt es ein majorisierendes Element z für M, d. h. $z \in R$, $z \geq \Theta$, $Mz > \Theta$.

Beweis: Mit $z = u_o$ gilt: $Mz = \hat{M}z - \sigma_o z + (p_o+\sigma_o)z = (p_o+\sigma_o)z > \Theta$.

BEMERKUNG 2.2: Satz 2.1 bleibt richtig, wenn in (V3)(a) $u_o \succ \Theta$ gefordert und in (V3)(b) nur $p_o(x) \geq -\sigma_o$ in $x_1 \leq x \leq x_2$ aber $p_o(x) \not\equiv -\sigma_o$ verlangt wird.
Bei den RB 1.3 gemäß Tabelle 1 mit $g_{o1} = g_{o2} = 0$ ist (V3)(a) erfüllt, da die EWA (5) den EW $\sigma_o = 0$ und die EF $u_o(x) \equiv 1$ besitzt. Hinreichend für (V3)(a) ist auch die Inverspositivität von $\hat{M} : R \to S$. Dazu wird zunächst gezeigt:

LEMMA 2.3: Der Operator $\hat{M} : R \to S$ sei inverspositiv und es sei $v_o \in \hat{Y} = C_4[x_1,x_2] \cap Y$, $v_o \succ \Theta$. Für die Lösung der RWA $\hat{M}v_1 = v_o$, $v_1 \in R$ gilt dann $v_1 \succ \Theta$.

Beweis: Es sei gesetzt für $n \in \mathbb{N}_o$: $\hat{G}^{(n)}(x,\xi) = \frac{d^n}{dx^n}\hat{G}(x,\xi)$.

Auf Grund der Inverspositivität von $\hat{M} : R \to S$ ist die zugehörige Greensche Funktion $\hat{G}$ nichtnegativ. Da $\hat{G}$ als Funktion von x die RB von R erfüllt, gilt dann $\hat{G}^{(n_1)}(x_1,\xi) \geq 0$ und $(-1)^{n_2}\hat{G}^{(n_2)}(x_2,\xi) \geq 0$ in $x_1 < \xi < x_2$. Mit dem in § 1 eingeführten Funktional l_x läßt sich dies zu $l_x\hat{G}(\cdot,\xi) \geq 0$ in $x_1 < \xi < x_2$ für $x \in [x_1,x_2]$ zusammenfassen. Wegen $v_1 = \hat{M}^{-1}v_o$ folgt

$l_x v_1 = \int_{x_1}^{x_2} l_x\hat{G}(\cdot,\xi)v_o(\xi)d\xi \geq 0$. Die Annahme $l_\zeta v_1 = 0$ für ein

$\zeta \in [x_1,x_2]$ führt wegen $v_o \succ \Theta$ zu $l_\zeta\hat{G}(\cdot,\xi) = 0$ in $x_1 < \xi < x_2$ und

wegen $v_o = \hat{M}^{-1}\hat{M}v_o$ zu $l_\zeta v_o = \int_{x_1}^{x_2} l_\zeta\hat{G}(\cdot,\xi)\hat{M}v_o(\xi)d\xi = 0$. Dies steht

im Widerspruch zu $l_\zeta v_o > 0$, was wegen $v_o \succ \Theta$ gelten muß. Damit gilt $l_x v_1 > 0$ für alle $x \in [x_1,x_2]$, d. h. $v_1 \succ \Theta$.

<u>SATZ 2.4:</u> Ist der Operator $\hat{M} : R \to S$ inverspositiv, so hat die EWA (5) einen positiven EW σ_o mit einer EF $u_o > \Theta$.

<u>Beweis:</u> Sei $v_o \in \hat{Y}$, $v_o \succ \Theta$; dann gilt nach Lemma 2.3 für die Lösung der RWA $\hat{M}v_1 = v_o$, $v_1 \in R$ ebenfalls $v_1 \succ \Theta$, so daß

$\mu = \inf\limits_{x_1<x<x_2} \frac{v_1(x)}{v_o(x)} > 0$ ausfällt und $\mu v_o \leq v_1 = \hat{M}^{-1}v_o$ erfüllt ist.

Da $\hat{M}^{-1}$ ein vollstetiger, positiver Operator auf S ist, besitzt $\hat{M}^{-1}$ nach Theorem 6.2 aus [8] einen positiven EW τ_o mit einer EF $u_o > \Theta$:

$$\tau_o u_o = \hat{M}^{-1}u_o \quad \text{bzw.} \quad \hat{M}u_o = \sigma_o u_o \quad \text{mit } \sigma_o = \frac{1}{\tau_o}\ .$$

<u>BEMERKUNG 2.5:</u> Nach Satz 4 aus [2] ist der Operator $\hat{M} : R \to S$ bei Inverspositivität sogar punktweise streng positiv, so daß für eine zu σ_o gehörige EF u_o gilt $u_o \succ \Theta$.

Die Inverspositivität von $\hat{M} : R \to S$ läßt sich auf die Inverspositivität bei Differentialoperatoren 2. Ordnung zurückführen. Dazu dienen die folgenden Bezeichnungen:

$$Lw = -(p_2w')' + p_1w$$

$$Bw(x_i) = w(x_i) \text{ oder } Bw(x_i) = g_{1i}w(x_i) + (-1)^i p_2w'(x_i) \quad (i=1,2) \tag{6}$$

$D = \{w \in C_2[x_1,x_2] \mid Bw(x_i) = 0 \ (i=1,2)\}$, $S = C_o[x_1,x_2]$

$D_1 = \{u \in C_2[x_1,x_2] \mid Bu(x_1) = 0,\ u(x_2) = 0\}$

$D_2 = \{v \in C_2[x_1,x_2] \mid v(x_1) = 0,\ Bv(x_2) = 0\}$

Ferner sei für jedes $t \in (x_1,x_2)$ gesetzt:

$D_1^t = \{u \in C_2[x_1,t] \mid Bu(x_1) = 0\}$ $\quad L_1^t u(x) = \begin{cases} Lu(x) & \text{in } x_1 \leq x < t \\ u(t) & \text{für } x = t \end{cases}$

$S_1^t = C_o[x_1,t]$

$D_2^t = \{v \in C_2[t,x_2] \mid Bv(x_2) = 0\}$ $\quad L_2^t v(x) = \begin{cases} v(t) & \text{für } x = t \\ Lv(t) & \text{in } t < x \leq x_2 \end{cases}$

$S_2^t = C_o[t,x_2]$

Die strenge Halbordnung in S_i^t $(i=1,2)$ sei mit $\succ$ bezeichnet.

Es gilt: $U \in S_1^t,\ U \succ \Theta \Leftrightarrow U(x) > 0$ in $x_1 \leq x \leq t$

$V \in S_2^t,\ V \succ \Theta \Leftrightarrow V(x) > 0$ in $t \leq x \leq x_2$.

Zur Charakterisierung der strengen Halbordnung in D_i^t $(i=1,2)$ werden die beiden Funktionale $l_{1,x}^t$ für $x \in [x_1,t]$ und $l_{2,x}^t$ für $x \in [t,x_2]$ eingeführt: Es sei $m_1 \in \mathbb{N}_o$ bzw. $m_2 \in \mathbb{N}_o$ die kleinste Ableitungsordnung, so daß für ein $u \in D_1^t$ bzw. ein $v \in D_2^t$ gilt $u^{(m_1)}(x_1) \neq 0$ bzw. $v^{(m_2)}(x_2) \neq 0$.

Sei $u \in D_1^t$: $\quad u \succ \Theta \Leftrightarrow l_{1,x}^t u > 0$ für alle $x_1 \leq x \leq t$

mit $l_{1,x_1}^t u = u^{(m_1)}(x_1)$, $l_{1,x}^t u = u(x)$ für $x_1 < x \leq t$.

Sei $v \in D_2^t$: $\quad v \succ \Theta \Leftrightarrow l_{2,x}^t v > 0$ für alle $t \leq x \leq x_2$

mit $l_{2,x}^t v = v(x)$ für $t \leq x < x_2$, $l_{2,x_2}^t v = (-1)^{m_2} v^{(m_2)}(x_2)$.

Zur Behandlung der RB 1.1 gemäß Tabelle 1 dient folgendes

<u>LEMMA 2.6:</u> Ist $L : D_i \to S$ inverspositiv, so ist auch $L_i^t : D_i^t \to S_i^t$ für jedes $t \in (x_1,x_2)$ inverspositiv $(i=1,2)$.

<u>Beweis:</u> OBdA sei i=1. Zum Nachweis der schwachen Präinverspositivität von L_1^t sei ein $u \in D_1^t$ mit $u \geq \Theta$, $L_1^t u \succ \Theta$, $u \not\succ \Theta$ vorgelegt. Der Fall $u(t) = 0$ steht im Widerspruch zu $u(t) = L_1^t u(t) > 0$. Im Fall $u(\xi) = 0$ für ein $\xi \in (x_1,t)$ folgt zunächst $u'(\xi) = 0$,

$u''(\xi) \geq 0$ und damit $L_1^t u(\xi) \leq 0$. Im Fall $u^{(m_1)}(x_1) = 0$ mit $0 \leq m_1 \leq 1$ gilt ebenfalls $u(x_1) = u'(x_1) = 0$, also $u''(x_1) \geq 0$ und somit $L_1^t u(x_1) \leq 0$. Beides steht im Widerspruch zu $L_1^t u \succ \Theta$. Zum Nachweis der Existenz eines stark majorisierenden Elementes für L_1^t wird zunächst die RWA $Lz = r$, $z \in D_1$ mit $r(x) \equiv 1$ betrachtet. Da $L : D_1 \to S$ inverspositiv ist, folgt $z \geq \Theta$. Es gilt sogar $z(x) > 0$ in $x_1 < x < x_2$, da L als inverspositiver Operator schwach präinverspositiv ist. Es sei z_1 die Einschränkung von z auf $[x_1,t]$. Dann gilt: $z_1 \in D_1^t$, $z_1 \geq \Theta$, $L_1^t z_1(x) = Lz(x) > 0$ in $x_1 \leq x < t$ und $L_1^t z_1(t) = z(t) > 0$, also $L_1^t z \succ \Theta$. Nach dem Monotoniesatz aus [9] folgt die Inverspositivität von L_1^t.

Fall	Voraussetzung (V4)	
1.1	$L : D_i \to S$ ist inverspositiv für i=1 und 2	
1.2.1	$g_{o2} = 0$	L : D → S ist inverspositiv
1.2.2	$g_{o1} = 0$	L : D → S ist inverspositiv
1.3.1	$g_{o1} > 0$, $g_{o2} = 0$	L : D → S ist inverspositiv
1.3.2	$g_{o1} = 0$, $g_{o2} > 0$	L : D → S ist inverspositiv

Tabelle 3

<u>SATZ 2.7:</u> Ist (V4) erfüllt, so ist $\hat{M} : R \to S$ inverspositiv, d. h. für alle $u \in R$ gilt: $\hat{M}u \geq \Theta \Rightarrow u \geq \Theta$.

<u>Beweis:</u> 1. Bei den RB 1.1 gemäß Tabelle 1 folgt aus (V4) nach Lemma 2.6, daß die Operatoren $L_i^t : D_i^t \to S_i^t$ für jedes $t \in (x_1,x_2)$ und $i \in \{1,2\}$ inverspositiv sind. Annahme: Ein $u \in R$ besitze in $t \in (x_1,x_2)$ ein negatives Minimum. Setzt man $w(x) = u'(x)$, so folgt $-\hat{M}u(x) = -(p_2w')''(x) + (p_1w)'(x) = (Lw)'(x)$ sowie $w(t) = u'(t) = 0$. Wegen $\hat{M}u(x) \geq 0$ in $x_1 \leq x \leq x_2$ ist Lw dort monoton nicht wachsend, hat also höchstens einen Vorzeichenwechsel in (x_1,x_2). Hat Lw keinen Vorzeichenwechsel in (x_1,x_2), so gilt entweder $Lw(x) \geq 0$ in $x_1 \leq x < t$ oder $Lw(x) \leq 0$ in $t < x \leq x_2$. Hat Lw einen Vorzeichenwechsel in (x_1,x_2), so gibt es ein $t_o \in (x_1,x_2)$ mit $Lw(t_o) = 0$. Es gilt dann

in $x_1 \le x \le t_o$: $0 \le \int_x^{t_o} \hat{M}u(s)ds = -Lw(t_o) + Lw(x) = Lw(x)$

in $t_o \le x \le x_2$: $0 \le \int_{t_o}^{x} \hat{M}u(s)ds = -Lw(x) + Lw(t_o) = -Lw(x)$.

Im Fall $t \le t_o$ folgt damit $Lw(x) \ge 0$ in $x_1 \le x < t$ und im Fall $t_o \le t$ folgt $Lw(x) \le 0$ in $t < x \le x_2$. Aus $Lw(x) \ge 0$ in $x_1 \le x < t$, $w(t) = 0$, also $L_1^t w(x) \ge 0$ in $x_1 \le x \le t$ und der RB $Bw(x_1) = 0$ ergibt sich, da $L_1^t : D_1^t \to S_1^t$ inverspositiv ist, $w(x) \ge 0$ in $x_1 \le x \le t$ und damit $u(t) = u(t) - u(x_1) = \int_{x_1}^{t} w(s)ds \ge 0$ im Widerspruch zur Annahme. Aus $Lw(x) \le 0$ in $t < x \le x_2$, $w(t) = 0$, also $L_2^t w(x) \le 0$ in $t \le x \le x_2$ und der RB $Bw(x_2) = 0$ ergibt sich, da $L_2^t : D_2^t \to S_2^t$ inverspositiv ist, $w(x) \le 0$ in $t \le x \le x_2$ und damit $-u(t) = u(x_2) - u(t) = \int_t^{x_2} w(s)ds \le 0$, also $u(t) \ge 0$ im Widerspruch zur Annahme.

2. Bei den RB 1.2.1 und 1.3.1 gemäß Tabelle 1 und 3 gilt wegen $g_{o2} = 0$ mit $w(x) = u'(x)$ $B_o u(x_2) = Lw(x_2) = 0$. Damit folgt in $x_1 \le x \le x_2$: $0 \le \int_x^{x_2} \hat{M}u(s)ds = -Lw(x_2) + Lw(x) = Lw(x)$. Zusammen mit den RB $Bw(x_i) = 0$ $(i=1,2)$ ergibt sich wegen der Inverspositivität von $L : D \to S$ $w(x) \ge 0$ in $x_1 \le x \le x_2$ und damit

$u(x) - u(x_1) = \int_{x_1}^{x} w(s)ds \ge 0$ in $x_1 \le x \le x_2$. Bei den RB 1.2.1 gilt $u(x_1) = 0$, also $u(x) \ge 0$ in $x_1 \le x \le x_2$. Bei den RB 1.3.1 gilt $B_o u(x_1) = g_{o1}u(x_1) - Lw(x_1) = 0$, also $u(x) \ge u(x_1) = \frac{1}{g_{o1}} Lw(x_1) \ge 0$ und damit $u(x) \ge 0$ in $x_1 \le x \le x_2$.

3. Für die RB 1.2.2 und 1.3.2 gemäß Tabelle 1 und 3 kann der Beweis analog zum vorangehenden Fall geführt werden.

Aus Satz 2.1 in Verbindung mit den Sätzen 2.4 und 2.7 folgt:

<u>SATZ 2.8:</u> Im Fall der RB 1.3 gemäß Tabelle 1 gelte $g_{o1} = g_{o2} = 0$ oder es sei (V4) erfüllt. Gilt noch (V3)(b), so gibt es ein schwach majorisierendes Element z für M.

Der Monotoniesatz aus [9] wird im Anschluß noch zum Nachweis der Inverspositivität des unter (6) eingeführten Differentialoperators $L : D \to S$ benutzt. In S wird dieselbe strenge Halbordnung verwendet wie die, die in § 1 eingeführt wurde. Zur Charakterisierung der strengen Halbordnung in D wird das Funktional $\tilde{l}_x$ für $x \in [x_1,x_2]$ benutzt: Es sei $m_i \in \mathbb{N}_o$ die kleinste Ableitungsordnung, so daß für ein $w_i \in D$ gilt $w_i^{(m_i)}(x_i) \neq 0$ $(i=1,2)$. Sei $w \in D$: $w \succ \theta \Leftrightarrow \tilde{l}_x w > 0$ für alle $x \in [x_1,x_2]$ mit $\tilde{l}_{x_1} w = w^{(m_1)}(x_1)$, $\tilde{l}_x w = w(x)$ in $x_1 < x < x_2$, $\tilde{l}_{x_2} w = (-1)^{m_2} w^{(m_2)}(x_2)$. Der Nachweis der schwachen Präinverspositivität von L, d. h. für alle $w \in D$, $w \geq \theta$, $Lw \succ \theta \Rightarrow w \succ \theta$, verläuft analog zum ersten Teil des Beweises von Lemma 2.6. Die Existenz eines stark majorisierenden Elementes $z \in D$ mit $z \geq \theta$, $Lz \succ \theta$ ist nach [9] (vgl. S. 77) gesichert, wenn folgende Bedingungen gelten:

(V5)(a) Sei $\rho \in C_o[x_1,x_2]$, $\rho(x) > 0$ in $x_1 \leq x \leq x_2$. Die EWA

$$\tilde{L}w = -(p_2 w')' = \kappa\rho w, \quad w \in D \tag{7}$$

besitze einen EW κ_o mit einer EF $w_o \succ \theta$.

(V5)(b) Es gelte $\inf\limits_{x_1<x<x_2} \frac{p_1(x)}{\rho(x)} > -\kappa_o$.

Im Fall $g_{11} = g_{12} = 0$ ist (V5)(a) erfüllt, da die EWA (7) den EW $\kappa_o = 0$ mit der EF $w_o(x) \equiv 1$ hat. Hinreichend für (V5)(a) ist auch die Inverspositivität von $\tilde{L} : D \to S$. Unter den in Tabelle 4 in Abhängigkeit von den vorliegenden RB angegebenen Bedingungen existiert die zu $\tilde{L}$ gehörige Greensche Funktion H und ist nichtnegativ. Durch

$$TW(x) = \int_{x_1}^{x_2} H(x,\xi)\rho(\xi)W(\xi)d\xi, \quad W \in S$$

wird sogar ein punktweise streng positiver Operator $T : S \to D$

definiert. Nach Theorem 6.2 aus [8] besitzt die EWA (7) einen positiven EW κ_o mit einer EF $w_o \succ \Theta$.

$$\Phi_1(x) = \int_{x_1}^{x} \frac{dt}{p_2(t)}, \quad \Phi_2(x) = \int_{x}^{x_2} \frac{dt}{p_2(t)}; \; \gamma = \Phi_1(x_2) = \Phi_2(x_1)$$

RB	Bedingung für g_{1i}	$H(x,\xi)$ für $\xi \leq x$
$w(x_1)=0$ $w(x_2)=0$	-	$\frac{\Phi_2(x)\Phi_1(\xi)}{\gamma}$
$w(x_1)=0$ $g_{12}w(x_2)+p_2w'(x_2)=0$	$1+g_{12}\gamma > 0$	$\frac{(1+g_{12}\Phi_2(x))\Phi_1(\xi)}{1+g_{12}\gamma}$
$g_{11}w(x_1)-p_2w'(x_1)=0$ $w(x_2)=0$	$1+g_{11}\gamma > 0$	$\frac{\Phi_2(x)(1+g_{11}\Phi_1(\xi))}{1+g_{11}\gamma}$
$g_{11}w(x_1)-p_2w'(x_1)=0$ $g_{12}w(x_2)+p_2w'(x_2)=0$	$1+g_{11}\gamma>0, 1+g_{12}\gamma>0$ $g_{11}+g_{12}+g_{11}g_{12}\gamma>0$	$\frac{(1+g_{12}\Phi_2(x))(1+g_{11}\Phi_1(\xi))}{g_{11}+g_{12}+g_{11}g_{12}\gamma}$

Tabelle 4

Damit ergibt sich der

<u>SATZ 2.9:</u> Für die Koeffizienten g_{1i} (i=1,2) der RB von D gelte $g_{11} = g_{12} = 0$ oder es seien die Bedingungen aus Tabelle 4 erfüllt. Ferner gelte (V5)(b). Dann ist der Operator $L : D \to S$ inverspositiv.

<u>BEMERKUNG 2.10:</u> Im Spezialfall $\rho(x) = \frac{1}{p_2(x)}$ geht die EWA (7) durch die Transformation $\zeta = \int_{x_1}^{x} \frac{dt}{p_2(t)}$ (vgl. z. B. [4], S. 323) in eine EWA mit konstanten Koeffizienten über:
$-v''(\zeta) = \kappa v(\zeta)$ in $\zeta_1 \leq \zeta \leq \zeta_2$, $\tilde{B}v(\zeta_i) = 0$ (i=1,2)
mit $\tilde{B}v(\zeta_i) = v(\zeta_i)$ oder $\tilde{B}v(\zeta_i) = g_{1i}v(\zeta_i) + (-1)^i v'(\zeta_i)$,
wobei $\zeta_1 = 0$ und $\zeta_2 = \gamma$ gesetzt ist.

<u>BEISPIEL 2.11:</u> $\hat{M}y(x) = y^{IV}(x) + A((1-x)y'(x))'$ bei den RB $y(0) = y'(0) = y''(1) = y'''(1) = 0$ ist nach Satz 2.7 inverspositiv, wenn $Lw(x) = -w''(x) - A(1-x)w(x)$ bei den RB $w(0) = w'(1) = 0$

inverspositiv ist. Dies ist nach Satz 2.9 für $A < (\frac{\pi}{2})^2$ der Fall.

§ 3 Eigenwerteinschließung

Die Ergebnisse aus § 1 und § 2 werden zusammengefaßt in

SATZ 3.1 Es gelte (V1), (V2), (V3). Dann ist der Operator $A : S \to S$ gemäß (3) vollstetig und positiv.

Beweis: Nach Satz 1.3 in Verbindung mit den Sätzen 1.5.2 und 2.1 sind die Operatoren $M : R \to S$ und $PM : R \to Y$ inverspositiv. Ferner ist $PN : R \to Y$ nach Satz 1.5.1 ein positiver Operator, so daß nach Satz 1.2 und wegen $Av = M^{-1}Nv = (PM)^{-1}PNv$ für $v \in R$ die Einschränkung von A auf R ein positiver Operator ist. Damit ist die Kernfunktion $N_\xi G$ des Integraloperators A notwendig nichtnegativ, also auch A auf S ein positiver Operator. Wegen der Stetigkeit von $N_\xi G(x,\xi)$ in $x_1 \leq x,\xi \leq x_2$ ist A vollstetig als Abbildung des Banachraumes $S = C_o[x_1,x_2]$ in sich (vgl. z. B. [9], S. 114).

Es sei $K = \{v \in S | v \geq \Theta\}$ und $v_o \in Y$, $v_o \succ \Theta$. Wegen $A : S \to Y$ wird durch

$$l(v) = \sup_{x_1<x<x_2} \frac{Av(x)}{v_o(x)}, \quad v \in S \tag{8}$$

(vgl. [1]) ein reellwertiges Funktional auf S definiert, welches folgende Eigenschaften hat: $l(v) \geq 0$ für $v \in K$; $l(\alpha v) = \alpha l(v)$ für $\alpha \geq 0$, $v \in K$; $\Theta \leq u \leq v \Rightarrow l(u) \leq l(v)$; $Av \leq l(v)v_o$ für $v \in S$.

SATZ 3.2 Für die EWA (1) seien (V1), (V2), (V3) erfüllt; es gebe $v_o \in Y$, $v_1 \in R$ mit $v_o \succ \Theta$, $Mv_1 = Nv_o$.

$$\Phi_{min} = \inf_{x_1<x<x_2} \Phi(x), \quad \Phi_{max} = \sup_{x_1<x<x_2} \Phi(x) \text{ mit } \Phi(x) = \frac{v_o(x)}{v_1(x)}.$$

Gilt $\Phi_{max} < \infty$, so besitzt die EWA (1) einen positiven EW λ_o mit einer nichtnegativen EF; λ_o ist betragskleinster EW dieser EWA und es gilt $\Phi_{min} \leq \lambda_o \leq \Phi_{max}$.

Beweis: Nach Satz 3.1 ist A vollstetig und positiv auf S und es gilt $\mu v_o \leq v_1 = Av_o$ mit $\mu = \inf_{x_1<x<x_2} \frac{v_1(x)}{v_o(x)}$, wobei $\mu > 0$ ist. Nach

Theorem 6.2 aus [8] besitzt A einen positiven EW κ_o mit einer EF $u_o > \Theta$; κ_o ist betragsgrößter EW von A und es gilt $\mu \leq \kappa_o$. Aus $\kappa_o u_o = Au_o \leq l(u_o)v_o$ folgt durch Anwendung des Funktionals l gemäß (8) die Abschätzung: $\kappa_o \leq l(v_o)$. Wegen der Äquivalenz der EWAn (1) und (3) wird $\lambda_o = \frac{1}{\kappa_o}$ betragskleinster EW der EWA (1) und es gilt obige Einschließung.

BEMERKUNG 3.3: Bei der hier behandelten Klasse allgemeiner EWAn ist eine EW-Einschließung durch Infimum und Supremum nur eines Quotienten möglich, nämlich $\Phi_o(x) = \frac{v_o(x)}{v_1(x)}$; der Quotient $\Phi_1(x) = \frac{v_o'(x)}{v_1'(x)}$ braucht hier nicht berücksichtigt zu werden. Allgemein ist diese Aussage aber nicht gültig, wie das Gegenbeispiel in [6] zeigt.

BEMERKUNG 3.4: Neben dem Satz 3.2 ist ein in [5] angegebener Quotienteneinschließungssatz auf die hier untersuchte Klasse allgemeiner EWAn anwendbar: Ist y_s eine nichtnegative EF zum EW λ_s der EWA (1) und gibt es ein $w \in R$ mit $Nw(x) \geq 0$ in $x_1 \leq x \leq x_2$, so gilt, wenn $\Psi(x) = \frac{Mw(x)}{Nw(x)}$ in $x_1 < x < x_2$ die Bedingung $0 < \Psi(x) < \infty$ erfüllt, die EW-Einschließung

$$\Psi_{min} = \inf_{x_1<x<x_2} \Psi(x) \leq \lambda_s \leq \sup_{x_1<x<x_2} \Psi(x) = \Psi_{max}.$$

Für die Funktionen $v_o, v_1 \in R$ gelte $v_o \succ \Theta$, $Mv_1 = Nv_o$ und für $w = v_o$ oder $w = v_1$ sei $Nw(x) \geq 0$ in $x_1 \leq x \leq x_2$ erfüllt. Dann folgt zunächst für $x_1 \leq x \leq x_2$

$$\Psi_{min} \cdot Nw(x) \leq Mw(x) \leq \Psi_{max} \cdot Nw(x).$$

Im Fall $w = v_o$ folgt mit der Inverspositivität von M und wegen $v_1 = M^{-1}Nv_o$: $\Psi_{min} \leq \Phi(x) \leq \Psi_{max}$. Ist im Fall $w = v_1$ der Operator $N : Y \to S$ inverspositiv, so gilt dieselbe Abschätzung wegen $N^{-1}Mv_1 = v_o$. In beiden Fällen liefert der Satz aus [5] jedoch keine besseren Schranken als der Satz 3.2.

§ 4 Anwendungen und numerische Beispiele

Zunächst werden hinreichende Bedingungen für die Forderung

$p(x) \geq 0$ in $x_1 \leq x \leq x_2$ aus (V2) angegeben.

SATZ 4.1: Es gelte $q_1(x) > 0$ in $x_1 \leq x \leq x_2$. Die homogene RWA $Nw = \Theta$, $U_1[w] = V_1[w] = 0$ habe nur die triviale Lösung. H bezeichne die zugehörige Greensche Funktion. Werden ψ und ϕ als Lösungen der RWAn bestimmt:

$$N\psi(x) = s(x) \text{ in } x_1 \leq x \leq x_2, \quad U_1[\psi] = V_1[\psi] = 0$$
$$N\phi(x) = r(x) \text{ in } x_1 \leq x \leq x_2, \quad U_1[\phi] = V_1[\phi] = 0$$

mit $r,s \in C_0[x_1,x_2]$, so gilt für $p(x) = W(\phi,\psi)(x)$ die Darstellung mit $x_1 \leq \xi \leq x \leq \eta \leq x_2$:

$$p(x) = \frac{1}{q_1(x)} \int_{x_1}^{x} \left(\int_{x}^{x_2} H(\xi,\eta)[r(\xi)s(\eta)-r(\eta)s(\xi)]d\eta \right) d\xi .$$

Beweis: Die Greensche Funktion H läßt sich in Produktform angeben (vgl. z. B. [4], S. 328). Werden ϕ und ψ mit Hilfe von H dargestellt und in $p = W(\phi,\psi)$ eingesetzt, so ergibt sich die obige Formel.

FOLGERUNG 4.2: H sei nichtnegativ und für eine im Intervall $[x_1,x_2]$ stetige, positive Funktion α gelte:
$\alpha(x)s(x)$ sei in (x_1,x_2) monoton nicht fallend,
$\alpha(x)r(x)$ sei in (x_1,x_2) monoton nicht wachsend.
Dann folgt $p(x) \geq 0$ in $x_1 \leq x \leq x_2$.

Die nachfolgenden Beispiele sind aus [3] entnommen.

BEISPIEL 4.3: Elastisch gebetteter Druckstab.
$(\alpha y'')''+\kappa y = -\lambda y''$ in $[-1,1]$, $y(\pm 1) = y''(\pm 1) = 0$.
Dabei sei $\alpha(x) > 0$ in $-1 \leq x \leq 1$ und $\kappa \in \mathbb{R}$. Ist α eine gerade Funktion, so zerfällt die EWA in ein Teilproblem (G) mit geraden EF, welche die RB $y'(0) = y'''(0) = y(1) = y''(1) = 0$ erfüllen und in ein weiteres Teilproblem (U) mit ungeraden EF, welche die RB $y(0) = y''(0) = y(1) = y''(1) = 0$ erfüllen. Im Fall (G) kann die RB $y'''(0)$ wegen $\alpha'(0) = 0$ durch $(\alpha y'')'(0) = 0$ ersetzt werden.

Zum Nachweis von (V1), (V2) im Fall (G) werden ψ und ϕ als Lösungen folgender RWAn bestimmt:

$$-\alpha(x)\psi''(x) = x^n \quad \text{in } 0 \leq x \leq 1, \quad \psi'(0) = \psi(1) = 0$$
$$-\alpha(x)\phi''(x) = (1-x)^m \quad \text{in } 0 \leq x \leq 1, \quad \phi'(0) = \phi(1) = 0.$$

Für $n = 2$, $\kappa \leq \frac{2}{\psi(0)}$ gilt $(\alpha\psi")'(0) = 0$ und $M\psi \leq \Theta$.

Für $m = 3$, $\kappa \leq \frac{6}{-\phi'(1)}$ gilt $\phi"(1) = 0$ und $M\phi \leq \Theta$.

Im Fall (U) werden die RWAn für ψ und ϕ betrachtet:

$-\alpha(x)\psi"(x) = x^n$ in $0 \leq x \leq 1$, $\psi(0) = \psi(1) = 0$

$-\alpha(x)\phi"(x) = (1-x)^m$ in $0 \leq x \leq 1$, $\phi(0) = \phi(1) = 0$.

Für $n = 3$, $\kappa \leq \frac{6}{\psi'(0)}$ gilt $\psi"(0) = 0$ und $M\psi \leq \Theta$.

Für $m = 3$, $\kappa \leq \frac{6}{-\phi'(1)}$ gilt $\phi"(1) = 0$ und $M\phi \leq \Theta$.

In den Fällen (G) und (U) ergibt sich mit $\alpha(x)s(x) = x^n$, $\alpha(x)r(x) = (1-x)^m$ nach Folgerung 4.2 $p(x) \geq 0$ in $0 \leq x \leq 1$. Bedingung (V4), die für (V3)(a) hinreichend ist, ist in den Fällen (G) und (U) erfüllt, da $Lw = -(\alpha w')'$ bei den RB $w(0) = w'(1) = 0$ und $w'(0) = w(1) = 0$ inverspositiv ist. Im Spezialfall $\alpha(x) = 3-x^2$ ergeben sich als zulässige Bereiche für κ:

(G) $-2(\frac{\Pi}{2})^4 < \kappa \leq \min\ (61.878,\ 70.311)$

(U) $-2\Pi^4 < \kappa \leq \min\ (300.390, 341.979)$

Die groben unteren Grenzen für κ ergeben sich jeweils aus einem Vergleichsproblem zur EWA (5) mit konstanten Koeffizienten; Bedingung (V3)(b) ist erfüllt, da das Vergleichsproblem zu (5) bei der Abschätzung $3-x^2 \geq 2$ in $0 \leq x \leq 1$ eine untere Schranke für den EW σ_o von (5) liefert.

Bei $\kappa = 60$ wurden ausgehend von Polynomen als Ansatzfunktionen für v_o und v_1, die sämtliche RB erfüllen und in der Beziehung $Mv_1 = Nv_o$ stehen, nach Satz 3.2 EW-Einschließungen berechnet; bei der Rechnung mitgeführte freie Parameter wurden dazu benutzt, möglichst große untere Schranken und möglichst kleine obere Schranken zu ermitteln.

Fall	Anzahl freier Parameter	EW-Einschließung	Schranken-differenz
	0	$31.0714 \leq \lambda_o^G \leq 31.4378$	0.3664
G	1	$31.3421 \leq \lambda_o^G \leq 31.3568$	0.0147
	2	$31.3523 \leq \lambda_o^G \leq 31.3534$	0.0011

Fall	Anzahl freier Parameter	EW-Einschließung	Schranken-differenz
	0	$32.0833 \leq \lambda_0^U \leq 33.1696$	1.0863
U	1	$32.6756 \leq \lambda_0^U \leq 32.7215$	0.0459
	2	$32.7098 \leq \lambda_0^U \leq 32.7137$	0.0039

BEISPIEL 4.4: Die in geschlossener Form lösbare EWA

$y^{IV} + \kappa y = -\lambda y''$ in $[-1,1]$, $y(\pm 1) = y''(\pm 1) = 0$

mit $\kappa \in \mathbb{R}$ besitzt die EW und EF:

$$\lambda_n = \left(n\frac{\Pi}{2}\right)^2 + \frac{\kappa}{\left(n\frac{\Pi}{2}\right)^2},\ n \in \mathbb{N}; \qquad \begin{aligned} y_k^{(G)}(x) &= \cos\left((2k-1)\frac{\Pi}{2}x\right), & k \in \mathbb{N} \\ y_k^{(U)}(x) &= \sin\left(2k\frac{\Pi}{2}x\right), & k \in \mathbb{N}. \end{aligned}$$

Im Falle $\kappa > 4\left(\frac{\Pi}{2}\right)^4$ gehört die nichtnegative EF $y_1(x) = \cos\frac{\Pi}{2}x$ nicht zum betragskleinsten EW. Alle übrigen EF haben wenigstens einen Vorzeichenwechsel. Zur Anwendung des Reduktionsverfahrens wird die EWA wieder in zwei Teilprobleme mit den RB aufgeteilt: (G) $y'(0) = y'''(0) = y(1) = y''(1) = 0$, (U) $y(0) = y''(0) = y(1) = y''(1) = 0$. Da für die Differentialgleichung $My = \Theta$ in jedem der Fälle $\kappa > 0$, $\kappa = 0$, $\kappa < 0$ ein Fundamentalsystem bekannt ist, lassen sich Funktionen ψ und ϕ als nichttriviale Lösungen folgender RWAn gewinnen:

$\psi^{IV} + \kappa\psi = 0$ in $[0,1]$	$\phi^{IV} + \kappa\phi = 0$ in $[0,1]$
(G) $\psi'(0) = \psi'''(0) = \psi(1) = 0$	(G) $\phi'(0) = \phi(1) = \phi''(1) = 0$
(U) $\psi(0) = \psi''(0) = \psi(1) = 0$	(U) $\phi(0) = \phi(1) = \phi''(1) = 0$.

Die Voraussetzungen (V1), (V2), (V3) sind dann erfüllbar:

im Fall (G) für $-\left(\frac{\Pi}{2}\right)^4 < \kappa \leq 4\left(\frac{\Pi}{2}\right)^4 \cong 24.352$,

im Fall (U) für $-\Pi^4 < \kappa \leq 4\rho_1^4 \cong 125.136$.

Dabei ist ρ_1 die kleinste positive Lösung der Gleichung $\cosh\rho \cdot \sin\rho + \sinh\rho \cdot \cos\rho = 0$; $\rho_1 > 2.3650$. Gilt bei

(G) $4\left(\frac{\Pi}{2}\right)^4 < \kappa < 4\left(\frac{3\Pi}{2}\right)^4$ und bei (U) $4\rho_1^4 < \kappa < 4\left(\frac{3\Pi}{2}\right)^4$,

so nimmt $\phi''(x)$ bei $x = 1$ positive Werte an und (V2) ist verletzt.

Die in Beispiel 4.3 verwendeten Ansätze für ψ und ϕ führen zu folgenden oberen Grenzen für κ: (G): $\kappa \leq 24$; (U): $\kappa \leq 120$.

BEISPIEL 4.5: Knicken schwerer Gestänge.
$(\alpha y'')'' - (Gy')' = -\lambda y''$ in $[0,1]$, $y(0) = y''(0) = y(1) = y''(1) = 0$.
Speziell sei $\alpha(x) \equiv 1$, $G(x) = \gamma x$ mit $\gamma \in \mathbb{R}$.
Zum Nachweis von (V1), (V2) werden ψ und ϕ als Lösungen der RWAn bestimmt:

$$-\psi''(x) = x^n \quad \text{in } 0 \leq x \leq 1, \quad \psi(0) = \psi(1) = 0$$
$$-\phi''(x) = (1-x)^m \quad \text{in } 0 \leq x \leq 1, \quad \phi(0) = \phi(1) = 0.$$

Für $n = 2$, $-24 \leq \gamma \leq \frac{8}{5}$ oder $n > 2$, $0 \leq \gamma \leq n(n+2)\frac{n-1}{n+3}$ gilt $\psi''(0) = 0$, $M\psi \leq \theta$. Für $m = 2$, $-8 \leq \gamma \leq 12$ gilt $\phi''(1) = 0$, $M\phi \leq \theta$.
Nach Folgerung 4.2 ergibt sich wieder $p(x) \geq 0$ in $0 \leq x \leq 1$.
Voraussetzung (V4) ist für $\gamma > -(\frac{\pi}{2})^2$ erfüllt, da dann $Lw = -w'' + \gamma x w$ bei den RB $w'(0) = w(1) = 0$ und $w(0) = w'(1) = 0$ inverspositiv ist. Bedingung (V3)(a) ist damit gewährleistet; (V3)(b) ist erfüllt wegen $p_0(x) \equiv 0$. Bei $\gamma = 1$ ergeben sich folgende EW-Einschließungen:

Anzahl freier Parameter	EW-Einschließung	Schranken-differenz
0	$10.2930 \leq \lambda_0 \leq 10.6581$	0.3651
1	$10.3347 \leq \lambda_0 \leq 10.4968$	0.1621
2	$10.3656 \leq \lambda_0 \leq 10.3694$	0.0038

BEISPIEL 4.6: Knickstab mit elastischer Zwischenstütze.
$(\alpha y'')'' = -\lambda y''$ in $[0,1]$, $y(0) = y''(0) = y'(1) = cy(1) - (\alpha y'')'(1) = 0$.
Es sei $\alpha(x) > 0$ in $0 \leq x \leq 1$, $c \in \mathbb{R}$.
Zum Nachweis von (V1), (V2) werden ψ und ϕ als Lösungen der RWAn bestimmt:

$$-\alpha\psi''(x) = x^n \quad \text{in } 0 \leq x \leq 1, \quad \psi(0) = \psi'(1) = 0$$
$$-\alpha\phi''(x) = 1-x \quad \text{in } 0 \leq x \leq 1, \quad \phi(0) = \phi'(1) = 0.$$

Für $n \in \mathbb{N}$ gilt $\psi''(0) = 0$ und $M\psi(x) \leq 0$.

Für $c \leq \frac{1}{\phi(1)}$ mit $\phi(1) = \int_0^1 \frac{\xi(1-\xi)}{\alpha(\xi)}\,d\xi$ gilt $c\phi(1) - (\alpha\phi'')'(1) \leq 0$.

Ferner gilt $M\phi = \Theta$. Zum Nachweis der Existenz eines majorisierenden Elementes z für M wird die RWA $(\alpha z'')'' = 1$ in $[0,1]$,

$z(0) = z''(0) = z'(1) = cz(1) - (\alpha z'')'(1) = 0$ betrachtet. Die Lösung z der RWA läßt sich für $c \geq 0$ in der Form darstellen:

$$z(x) = \int_0^1 \left(\int_0^x \tilde{H}(s,t)\,ds\right)(1-t)\,dt \text{ mit } \tilde{H}(s,t) = H(s,t) - \frac{c}{1+c\eta}\int_0^1 H(s,\tau)\,d\tau \cdot$$

$\cdot \int_0^1 H(\sigma,t)\,d\sigma$, wobei H Greensche Funktion zu $Lw = -(\alpha w')'$, $w'(0) = w(1) = 0$ ist und $\eta = \int_0^1 \int_0^1 H(s,t)\,ds\,dt$. Für $c \leq \frac{1}{\phi(1)}$ gilt $\tilde{H}(s,t) \geq 0$ in $0 \leq s,t \leq 1$ (vgl. [7], S. 99).

Literaturverzeichnis

[1] BOHL, E.: Die Theorie einer Klasse linearer Operatoren und Existenzsätze für Lösungen nichtlinearer Probleme in halbgeordneten Banachräumen. Arch. Rational Mech. Anal. 15 (1964), 263 - 288

[2] BOHL, E., BEYN, W. J. und LORENZ, J.: Zur Anwendung der Theorie über den Spektralradius linearer streng-monotoner Operatoren. ISNM 24 (1974), 23 - 31

[3] COLLATZ, L.: Eigenwertaufgaben mit technischen Anwendungen. Leipzig: Akad. Verlagsges. Geest & Portig K.G. 1963

[4] HARTMAN, P.: Ordinary differential equations. New York, London, Sydney: John Wiley & Sons, Inc. 1964

[5] HELD, W.: Die Collatzschen Einschließungssätze für Eigenwerte bei Differentialgleichungen. ISNM 24 (1974), 47 - 55

[6] HELD, W.: Bemerkungen zu einem Einschließungssatz von L. Collatz. ISNM 26 (1975), 63 - 66

[7] KLEIN, P. P.: Zur Einschließung des betragskleinsten Eigenwertes bei Eigenwertaufgaben mit Differentialgleichungen vierter Ordnung. ISNM 48 (1979), 85 - 101

[8] KREIN, M. G. und RUTMAN, M. A.: Linear operators leaving invariant a cone in a Banach space. Am. Math. Soc. Transl. Series One Vol. 10 (1962), 199 - 325

[9] SCHRÖDER, J.: Operator inequalities. New York: Academic Press 1980

Dr. Peter P. Klein
Rechenzentrum der TU Clausthal
Erzstraße 51
D-3392 Clausthal-Zellerfeld

ISNM, Vol. 69
Numerical Treatment of
Eigenvalue Problems, Vol. 3

APROXIMATION OF SOLUTIONS TO NONLINEAR EQUATIONS

Peter D. Miletta Geldorf.

In this paper we consider the appoximation of solutions to operator equations of the form

$$Au + N(u) = 0 \tag{1}$$

in a normed linear space with inner product by the Rayleigh-Ritz-Galerkin method. Here A is linear and N nonlinear. In particular we consider two specific questions associated with this approximation :

(1) When the existence of the Rayleigh-Ritz-Galerkin approximation implies the existence of a solution to (1)

(2) The calculation of error bounds on the approximation.

The method used to answer these questions is a varient of the Lyapunov-Schmidt or the Alternative method of Cesari.

Let B be a Banach Space with norm denoted by $\|u\|$ in which there is also defined a real inner product which will be denoted by (u,v). It will not be assumed that the norm in B is that induced by the inner product. However it will be assumed that A is self adjant with respect to (u,v), ie. $(Au,v) = (u, Av)$ for all $u, v \in D_A$.

On B we consider the nonlinear equation (1) where A is a linear operator with domain D_A and N nonlinear mapping its domain D_N into B. The sum A + N is defined on $D_A \cap D_B$ which is assumed to be dense in B. It will be assumed further that A has an infinite system of eigenvectors $A\varphi_i = \lambda_i\varphi_i$ with $\lambda_i \neq 0$, $i = 1, 2, \ldots$ and $|\lambda_i| \to \infty$ as $i \to \infty$. We assume that these eigenvectors are orthonormal with respect to the inner product. Each element $u \in B$ will have an "expansion" in terms of $\{\varphi_i\}$ given by

$$u \approx \sum_{i=1}^{\infty} (u, \varphi_i)\varphi_i$$

It will be assumed that this "expansion" converges in the usual sense with respect to the norm on B on a set containing $D_A \cap D_B$.

We let P^k denote the projection on the subspace of B spanned by the k vectors $\{\varphi_1, \ldots , \varphi_k\}$, that is

$$P^k u = \sum_{i=1}^{k} (u, \varphi_i)\varphi_i$$

We then approximate solutions to (1) by the Rayleigh-Ritz-Galerkin equation

$$P^k A P^k u + P^k N(P^k u) = 0 \qquad (2)$$

defined on the subspace $P^k B$.

In many applications it is found that the operator N is reproducing with respect to $\{\varphi_i\}$. That is $N(\sum_{i=1}^{n} \alpha_i \varphi_i) = \sum_{i=1}^{m(n)} \beta_i^n(\alpha_1, \ldots, \alpha_n)\varphi_i$ for $i = 1, 2, \ldots$, where the β_i^n's are assumed to be explicitely known functions of the α_y's and m(n) is some number which depends soley on n. In this case equation (2) takes on a simplier form. Letting $u_i = (u, \varphi_i)$, then equation (2) reduces to

$$\sum_{i=1}^{k} \lambda_i u_i \varphi_i + \sum_{i=1}^{k} \beta_i^k (u_1, \ldots, u_k)\varphi_i = 0 \qquad (3)$$

which we write in the equivalent form as a set of nonlinear equations

$$\lambda_i u_i + \beta_i^k(u_1, \ldots, u_k) = 0 \qquad (4)$$

$i = 1, 2, \ldots, k$.

Equation (2), or in the case of N reproducing with respect to $\{\varphi_i\}$ equations (3) or (4) will be called the kth determining equation for the u_i and a solution to these denoted by $u_1^k, \ldots, u_k^k$ will determine the kth order approximation

$$u^k = \sum_{i=1}^{k} u_i^k \varphi_i \qquad (5)$$

to a solution u of (1). We will assume that the kth determining equation has a solution $u_1^k, \ldots, u_k^k$ and will consider two important questions related to the approximation (5).

(1) Does the existence of an approximation u^k imply the existence of a solution to (1) ?

(2) If a solution to (1) does exist, can we compute error bounds on the approximation (5) ?

The method that will be used to answer these questions is a varient of the Lyapunov Schmidt method or the alternative method of Cesari. The starting point is the similarity between equation (2) and the equation

$$P^k(Ay + N(y)) = 0 \qquad (6)$$

In fact we will show that a certain mapping associated with the operator equation (1) has a fixed point in a neighborhood of u^k, that under certain circumstances this neighborhood can be chosen so that the fixed point satisfies (6) and becomes a solution to (1), and finally that the dimensions of the neighborhood can be measured so as to obtain error bounds on the approximation.

The results contained in this paper are a synthesis of results previously obtained by the author in [1], and [2], and the reader is refered to these articles for proofs of the theorems.

The first step is to construct an appropriate neighborhood of u^k and the mapping associated with (1). On B we let H denote the operator

$$Hu = \sum_{i=1}^{\infty} \frac{(u, \varphi_i)\varphi_i}{\lambda_i}$$

Lemma 1. The operator H is linear, bounded, commutes with P^k for all k, and moreover satisfies

$$P^k H(I - P^k) = 0$$

for all k.

Furthermore we note that $H(I - P^k)$ is bounded and we set $B_k = \| H(I-P^k) \|$. We let $B_A = \{u | u \in B, \|u\| \leq A\}$ where A is chosen so that $\|u^k\| \leq A_0 < A$ for some appropriate A_0. We set

$$F^k u = H(I - P^k)N(u)$$

and

$$J^k u = P^k u - F^k u$$

mapping B_A into B. We chose the numbers $c_1^k, \dots, c_k^k$, d^k so that $c_1^k > 0$, $\sum_{i=1}^{k} c_1^k < d^k$ and let $b_1^k, \dots, b_k^k$ be any k numbers such that $|b_i^k - u_i^k| \leq c_1^k$ for $i = 1, \dots, k$. We then define $B_A^k = B_A^k(b_1^k, \dots, b_k^k) = \{u | u \in B_A^k,\ P^k u = \bar{u}^k, \|u - u^k\| \leq d^k\}$ where $\bar{u}^k = \sum_{i=1}^{k} b_i^k \varphi_i$. The set B_A^k is nomempty as it contains $\bar{u}^k$ and is furthermore closed. Finally we set $\gamma^k = \max \|H(I - P^k)N(u)\|$ for $u \in B_A^k$, $Q^k = \|u^k\|$. We observe that if $Q^k + d^k \leq A$ then $B_A^k \subset B_A$. Finally on B_A we assume that N satisfies a Lipschitz condition, that is; there exists a constant L_A such that

$$\|N(u) - N(v)\| \leq L_A \|u-v\|$$

for all $u,v \in B_A$.

With these preliminaries we prove the following important theorem

Theorem 1 Suppose $B_k L_A < 1$, $Q^k + d^k \leq A$ and more over that $\sum_{i=1}^{k} c_i^k + \gamma^k < (1 - B_k L_A)d^k$. Then $J^k : B_A^k \to B_A^k$ is a contraction and has a fixed point y. The fixed point satisfies $\|g - \bar{u}^k\| \leq d^k - \sum_{i=1}^{k} c_L^k$.

Proof : The proof is identical to the proof of Theorem 1 in [2].

Once we have a fixed point $y \in B_A^k$, we observe that by operating on

$$y = P^k y - H(I - P^k)N(y)$$

by A we obtain

$$Ay = P^k Ay - (I - P^k)N(y) \qquad (6)$$

or

$$Ay + N(y) = P^k(Ay + N(y))$$

Using the language of Cesari we have reduced the problem of the existence of a solution to (1) to the alternative problem of the existence of a $y \in B_A^k$ which satisfies

$$\text{a) } y = P^k y - H(I - P^k)N(y) \qquad (7)$$

and

$$\text{b) } P^k(Ay + N(y)) = 0 \qquad (8)$$

The existence of a solution to (8) is given by the above theorem. The alternative problem then involves showing that we may choose the constants b_1^k , ..., b_k^k so that (8) is satisfied.

To show that we may choose the constants so as to satisfy (8) we make the following considerations.

Since $y = \sum_{i=1}^{k} b_i^k \varphi_i + \ldots$ we see that (8) is equivalent to showing that the system of equations

$$\lambda_i b_i^k + (N(\sum_{j=1}^{k} b_j^k \varphi_j + \ldots) \varphi_i) = 0 \qquad (9)$$

has a solution $b_1^k, \ldots, b_k^k$ which is in some sense close to the solution $u_i^k, \ldots, u_k^k$ of the determining equations.To do this, we set for $(\alpha_1, .., \alpha_k) \in R^k$

$$\beta_1^k (\alpha_1, \ldots, \alpha_k) = (N(\sum_{j=1}^{k} \alpha_j \varphi_j), \varphi_i)$$

and

$$\beta^k (\alpha_1, \ldots, \alpha_k) = (N(\sum_{j=1}^{k} \alpha_j \varphi_j + \ldots), \varphi_i)$$

for i = 1, 2, ..., k where $\sum_{j=1}^{k} \alpha_j \varphi_j + \ldots$ represents the fixed point whose existence is given in Theorem I . We observe that

$$|\beta_i^k(\alpha_1, \ldots, \alpha_k) - \beta_i^k(\alpha_1, \ldots, \alpha_k)| \leq L_A(B_k L_A d^k + \gamma^k)$$

and we consider the mapping of R^k into R^k given coordinate wise by

$$t_1^k\ (\alpha_1, \ldots, \alpha_k) = \lambda_i \alpha_i + \beta_i^k\ (\alpha_1, \ldots, \alpha_k)$$

$$T_1^k\ (\alpha_1, \ldots, \alpha_k) = \lambda_i \alpha_i + \beta_i^k\ (\alpha_1, \ldots, \alpha_k)$$

We may think of T^k as a perturbation of t^k in the sence that $T^k = t^k+(T^k-t^k)$ with $\|(T^k - t^k)_i\| \leqslant L_A(B_k L_A d^k + \gamma^k)$ for an appropriate A. Since t^k has a zero in $(u_1^k, \ldots, u_k^k)$, if it maps a neighborhood of $(u_1^k, \ldots, u_k^k)$ onto a large enough neighborhood of the origin then T^k will have a zero in the neighborhood of $(u_1^k, \ldots, u_k^k)$. What is large enough is made precise by the following.

We consider the closed neighborhood $N = \{(u_1^k + c_1, \ldots, u_k^k + c_k) \mid |c_i| \leqslant c_i^k, i = 1, \ldots, k\}$ about $(u_1^k), \ldots u_k^k)$ determined by the constants $c_1^k, \ldots, c_k^k$. If it is possible to choose these constants so that $t^k(N)$ contains the closed neighborhood. $D = \{(x_1,\ldots,x_k) \mid |x_i| \leqslant L_A(B_k L_A d^k \gamma^k)\}$ then $T^k(N)$ will contain the origen. That is, the system of equations (9) will have a solution in N. Thus :

<u>Theorem II</u> : Given the abore assumptions, if $t^k(N) \supset D$ then the system of equations (9) will have a solution $(\bar{u}_1^k, \ldots, \bar{u}_k^k)$ which satisfies $|u_i^k - \bar{u}_i^k| \leqslant c_i^k$.

<u>Corollary</u> : Under the above assumptions and those of Theorem 1. The fixed point y is a solution to (1) and moreover we have $|u_i^k - \bar{u}_i^k| \leqslant c_i^k$; $\|y - u^k\| \leqslant d^k$; and $\|y - \bar{u}^k\| \leqslant B_k L_A d^k + \gamma^k$ where u^k is the kth approximation given by the Rayleigh-Ritz-Galerkin method of approximation.

The point of this theorem and its corollary is that a solution $u_1^k, \ldots, u_k^k$ to the determining equations determines the existence of a solution y to (1) and error bounds (the constants $c_1^k, \ldots, c_k^k, d^k$) on the approximation to y given by $u^k = \sum_{i=1}^k u_1^k \varphi_i$ essentially if the constants $c_1^k, \ldots, c_k^k, d^k$ can be chosen so that the image of the rectangular region N determined by these constants contains the cube D. In general it is very difficult to determine the c_i^k s (one usually sets $d^k=(1-B_k L_A)^{-1}(\sum_{i=1}^k c_i^k + \gamma^k))$ However when N is reproducing with respect to the $\{\varphi_i\}$ a very simple geometric argument can be used that leads to the calculation of the constants. The essential aspect of this method is that once the functions $\beta_i^k(\alpha_1,\ldots,\alpha_k)$ are explicitely known the image of the boundary of N can be determined. Specifically we consider the map t^k on the faces of the rectangle N for small values of the constants c_i. If these constants can be chosen so that t_i^k will map one of the faces of the rectangle perpendicular to the $i^{\underline{th}}$ axis greater than zero and the other such face less than zero, then we try to choose the constants so that the minimum value of u_i^k on the first face is greater than $L_A(B_k L_A d^k + \gamma^k)$ and the maximun value on the second face is less than $-L_A(B_k L_A d^k + \gamma^k)$. This procedure leads to a system of 2k inequalities (which generally reduces to a system of k inequalities) for the determination of the c_i^k,s. Complete details of this method is given in [1].

As an application of this theory we consider the problem of the existence and approximation of periodic solutions to second order ordinary differential equations of the form

$$x'' + g(t,x) = 0$$

Here it is assumed that g is defined for $|x| \leq A, |t| < \infty$, is $T = \frac{2\Pi}{\omega}$ periodic, and for each x is integrable on $[0,T]$. Moreover we assume that g satisfies the symmetry conditions

$$g(-t,-x) = -g(t,x) \; ; \; g(\tfrac{T}{4} + t,x) = g(\tfrac{T}{4} - t,x)$$

Finally we assume that g is uniformly Lipschitzian with respect to x with Lipschitz constant L. It is shown in [_] that we may consider this problem in the space B of all continuous $T = \frac{2\Pi}{\omega}$ periodic functions which satisfy the symmetry conditions

$$x(t) = -x(-t) \; ; \; x(\tfrac{T}{2} - t) = -x(\tfrac{T}{2} + t)$$

with norm $\|x\| = \max_{0 \leq t \leq T} |x(t)|$, the inner product in this space is taken as $(f_1,f_2) = \frac{2}{T} \int_0^T f_1(t) \cdot f_2(t)dt$ With $A = \frac{d^2}{dt^2}$ it is found $\{\sin(2_j-1)\omega t \mid |y = 1,2,...\}$ is a set of eigenvectors and more over each twice continuously differential function in B will have a Fourier series expansion in terms of the odd sine functions converging in norm. Finally the B_k is may be caclulated numerically and we find

$$B_1 \approx .\frac{14454}{\omega^2} \; , \; B_2 \approx .\frac{05524}{\omega^2} \; , \; B_3 \approx \frac{02952}{\omega^2}$$

In this spaces the Rayleigh-Ritz-Galerkin method is identical with the method of Harmonic Balance.

Taking $N(x) = g(t,x) = x + \frac{1}{2}x^3 - \frac{1}{2}\sin\frac{3}{2}t$ we consider the Duffing equation

$$x'' + x + \frac{1}{2}x^3 = \frac{1}{2}\sin\frac{3}{2}t.$$

Rutkowski [3] has shown that the nonlinearity is reproducing with respect to the odd sine functions and his calculated explicity the $B^k_{i\infty}$ is. It is shown in [2] that this equation has a periodic solution $x(t) = \sum_{i=1}^{\infty} u_1 \sin \frac{3}{2}(2i-1)t$ That is approximated by $x^3 = 1.971454 \sin \frac{3}{2} - .05879434 \sin \frac{9}{2} t +$ $.00168616 \sin \frac{15}{2} t$. Error bounds to this approximation were calculated and it was found that.

$|u_1 - 1.971454| \leqslant .0006994$; $|u_2 + .05879434| \leqslant .00030503$,

$|u_3 - .00168616| \leqslant .00009440$ and more over

$\| x(t) - x^3(t) \| \leqslant .00138743$

References

[1] Miletta, P : Existence and Approximation of Solutions to Non linear Operator Equations. Unpublished Manuscript.

[2] Miletta, P : Approximation of Periodic Solutions to Second Order Equations. Unpublished Manuscript.

[3] Rutkowski, P. : Approximate Solutions of Eigenvalue Problems with Reproducing Nonlinearities. University of Cologne, Appl. Math. Report 82-04 (1982).

Peter D. Miletta Geldorf
Depto. de Matemática
Universidad de Santiago de Chile
Casilla 5659
Santiago 2 Chile.

ISNM, Vol. 69
Numerical Treatment of
Eigenvalue Problems, Vol. 3

AN ELEMENTARY PROOF OF MONOTONY OF THE TEMPLE QUOTIENTS

Karel Rektorys

Monotony of the Temple quotients has been proved recently by F. Goerisch and J. Albrecht in their common work [1]. In the present paper, another proof - let us call it an elementary one - of this fact is presented.

Consider the eigenvalue problem

(1) $M\phi = \lambda N\phi$,

where the operators M and N satisfy standard assumptions of symmetry and positive definiteness on their domains of definition $D(M) \subset D(N) \subset H$, where H is a Hilbert space with the scalar product $\langle \cdot, \cdot \rangle$. For details see [1]. (Or the bilinear forms, corresponding to the operators M, N, fulfil the standard assumptions of symmetry and V-ellipticity; for details see [4].) Then there exists a system of positive eigenvalues

(2) $0 < \lambda_1 \leqq \lambda_2 \leqq \lambda_3 \leqq \ldots, \lim_{j\to\infty} \lambda_j = +\infty$

and a complete system of corresponding eigenelements (eigenfunctions)

(3) $\phi_1, \phi_2, \phi_3, \ldots$,

orthonormalized in the sense $\langle \phi_i, N\phi_j \rangle = \delta_{ij}$. (The finite dimensional case is not excluded, being not of such interest, of course.)

Let the least eigenvalue λ_1 be simple. When the Collatz method is applied to find twosided estimates for λ_1, one chooses an element $F_0 \in D(N)$ and finds, subsequently, the elements $F_j \in D(M)$, $j = 1, 2, 3, \ldots$, satisfying $MF_j = NF_{j-1}$. Let, moreover, F_0 and F_1 be linearly independent.

Construct, successively, the so-called Schwarz coefficients, Schwarz quotients and Temple quotients by

(4) $$a_j = \langle F_0, NF_j \rangle > 0,\ j = 0, 1, 2, \ldots,$$

(5) $$\mu_j = \frac{a_{j-1}}{a_j},\ j = 1, 2, 3, \ldots,$$

(6) $$\tau_j(L) = \frac{La_j - a_{j-1}}{La_{j+1} - a_j};\ \mu_1 < L < \lambda_2 .$$

It is well known that

(7) $$\tau_j(L) \leqq \lambda_1 \leqq \mu_j ,$$

(8) $$\mu_1 > \mu_2 > \mu_3 > \ldots .$$

We are going to prove that

(9) $$\tau_1(L) < \tau_2(L) < \tau_3(L) < \ldots \text{ for every } L \in (\mu_1, \lambda_2).$$

(As announced above, this monotony property of the Temple quotients has been proved by F. Goerisch and J. Albrecht. We only present here a proof based on a quite different idea.)

Thus let j be a fixed - otherwise arbitrary - positive integer. We have to prove that

(10) $$\tau_{j+1}(L) - \tau_j(L) > 0 \text{ for every } L \in (\mu_1, \lambda_2).$$

One has

(11) $$\tau_{j+1}(L) - \tau_j(L) = \frac{La_{j+1} - a_j}{La_{j+2} - a_{j+1}} - \frac{La_j - a_{j-1}}{La_{j+1} - a_j} = \frac{y_j(L)}{(La_{j+2} - a_{j+1})(La_{j+1} - a_j)}$$

with

(12) $$y_j(L) = (a_{j+1}^2 - a_j a_{j+2})L^2 + (a_{j-1}a_{j+2} - a_j a_{j+1})L + (a_j^2 - a_{j-1}a_{j+1}).$$

Now,

$$La_{j+2} - a_{j+1} = a_{j+2}(L - \mu_{j+2}) > 0 ,$$

$$La_{j+1} - a_j \quad = a_{j+1}(L - \mu_{j+1}) > 0$$

in virtue of (8) and of the fact that $L > \mu_1$. So to prove (10) we have to show that

(13) $\qquad y_j(L) > 0$ for every $L \in (\mu_1, \lambda_2)$.

The function y_j is defined by (12) for all (real) L. Because

$$a_{j+1}^2 - a_j a_{j+2} = a_{j+1}^2(1 - \frac{\mu_{j+1}}{\mu_{j+2}}) < 0 ,$$

this function is strictly concave everywhere. Thus to prove (13), it is sufficient to find two points L_1, L_2 satisfying

(14) $\qquad L_1 \leqq \mu_1, \ L_2 \geqq \lambda_2$

and, at the same time,

(15) $\qquad y_j(L_1) \geqq 0, \ y_j(L_2) \geqq 0 .$

j being fixed, let us choose

(16) $\qquad L_1 = \mu_j, \ L_2 = \dfrac{\lambda_1 a_j - a_{j-1}}{\lambda_1 a_{j+1} - a_j} .$

The first of the requirements (14) is satisfied by (8), the second one follows from [3], ineqs. (2.38), (2.39), p. 223.

An easy computation which we are not going to carry out here (for details see [4]) yields

(17) $\qquad y_j(\mu_j) = \dfrac{(a_{j-1}a_{j+1} - a_j^2)^2}{a_j^2} \geqq 0 ,$

(18) $\qquad y_j\left(\dfrac{\lambda_1 a_j - a_{j-1}}{\lambda_1 a_{j+1} - a_j}\right) = \dfrac{a_j^2 - a_{j-1}a_{j+1}}{(\lambda_1 a_{j+1} - a_j)^2} y_j(\lambda_1) .$

By (17), the first of the inequalities (15) is fulfilled. Because $a_j^2 - a_{j-1}a_{j+1} < 0$ by (8), we have to prove that

(19) $\qquad y_j(\lambda_1) \leqq 0$

to establish validity of the second one.

By definition ((11), (12)),

$$(20)\quad y_j(\lambda_1)=(\lambda_1 a_{j+2}-a_{j+1})(\lambda_1 a_{j+1}-a_j)\left(\frac{\lambda_1 a_{j+1}-a_j}{\lambda_1 a_{j+2}-a_{j+1}}-\frac{\lambda_1 a_j-a_{j-1}}{\lambda_1 a_{j+1}-a_j}\right).$$

The expressions in the first two brackets on the right-hand side of (20) are negative because of $\lambda_1 < \mu_{j+1}$, $\lambda_1 < \mu_{j+2}$. Thus to prove (19) it is sufficient to prove that

$$\frac{\lambda_1 a_{j+1} - a_j}{\lambda_1 a_{j+2} - a_{j+1}} - \frac{\lambda_1 a_j - a_{j-1}}{\lambda_1 a_{j+1} - a_j} \leqq 0 ,$$

or, each of these fractions being positive, that

$$(21)\qquad Q = \frac{\dfrac{\lambda_1 a_{j+1} - a_j}{\lambda_1 a_{j+2} - a_{j+1}}}{\dfrac{\lambda_1 a_j - a_{j-1}}{\lambda_1 a_{j+1} - a_j}} = \frac{\dfrac{a_j - \lambda_1 a_{j+1}}{a_{j+1} - \lambda_1 a_{j+2}}}{\dfrac{a_{j-1} - \lambda_1 a_j}{a_j - \lambda_1 a_{j+1}}} \leqq 1 .$$

To prove (21), the relation

$$(22)\qquad a_j = \sum_{i=1}^{\infty} \frac{\alpha_i^2}{\lambda_i^j}, \text{ where } \alpha_i = \langle F_0, N\phi_i \rangle,\ i = 1, 2, 3, \ldots$$

(for ϕ_i see (3)) will be applied, proved in [3] ((2.26), p. 221). Because

$$\sum_{i=1}^{\infty} \alpha_i^2 < \infty \quad \text{and} \quad \lambda_i \to +\infty \text{ for } i \to \infty,$$

we have

$$(23)\qquad Q = \lim_{p\to\infty} Q_p ,$$

where Q_p is obtained from Q if in (22) only p first summands are taken, i.e.

$$(24)\qquad Q = \frac{\dfrac{\sum_{i=1}^{p} \frac{\alpha_i^2}{\lambda_i^j} - \lambda_1 \sum_{i=1}^{p} \frac{\alpha_i^2}{\lambda_i^{j+1}}}{\sum_{i=1}^{p} \frac{\alpha_i^2}{\lambda_i^{j+1}} - \lambda_1 \sum_{i=1}^{p} \frac{\alpha_i^2}{\lambda_i^{j+2}}}}{\dfrac{\sum_{i=1}^{p} \frac{\alpha_i^2}{\lambda_i^{j-1}} - \lambda_1 \sum_{i=1}^{p} \frac{\alpha_i^2}{\lambda_i^{j}}}{\sum_{i=1}^{p} \frac{\alpha_i^2}{\lambda_i^{j}} - \lambda_1 \sum_{i=1}^{p} \frac{\alpha_i^2}{\lambda_i^{j+1}}}} .$$

Note that, making the differences, the first summands drop out : $\alpha_1^2/\lambda_1^j - \lambda_1 \alpha_1^2/\lambda_1^{j+1} = 0$, etc. Thus, after a simple rearranging,

$$\sum_{i=1}^{p} \frac{\alpha_i^2}{\lambda_i^j} - \lambda_1 \sum_{i=1}^{p} \frac{\alpha_i^2}{\lambda_i^{j+1}} = \frac{\alpha_2^2}{\lambda_2^j}(1 - \frac{\lambda_1}{\lambda_2}) + \frac{\alpha_3^2}{\lambda_3^j}(1 - \frac{\lambda_1}{\lambda_3}) + \ldots$$

$$\ldots + \frac{\alpha_p^2}{\lambda_p^j}(1 - \frac{\lambda_1}{\lambda_p}) = \beta_2 \nu_2^j + \beta_3 \nu_3^j + \ldots + \beta_p \nu_p^j ,$$

where, for simplicity, the notation

$$\alpha_i^2(1 - \frac{\lambda_1}{\lambda_i}) = \beta_i \geqq 0, \qquad \nu_i = \frac{1}{\lambda_i} > 0$$

has been used, etc. Thus, we get, finally,

$$(25)\quad Q_p = \frac{\dfrac{\sum_{i=2}^{p} \beta_i \nu_i^j}{\sum_{i=2}^{p} \beta_i \nu_i^{j+1}}}{\dfrac{\sum_{i=2}^{p} \beta_i \nu_i^{j-1}}{\sum_{i=2}^{p} \beta_i \nu_i^{j}}} = \frac{\sum_{i=2}^{p} \beta_i^2 \nu_i^{2j} + 2 \sum_{\substack{i,k=2 \\ i<k}}^{p} \beta_i \beta_k \nu_i^j \nu_k^j}{\sum_{i=2}^{p} \beta_i^2 \nu_i^{2j} + \sum_{\substack{i,k=2 \\ i<k}}^{p} \beta_i \beta_k (\nu_i^{j+1} \nu_k^{j-1} + \nu_i^{j-1} \nu_k^{j+1})} \leqq 1,$$

because

$2\nu_i^j \nu_k^j = \nu_i^{j-1} \nu_k^{j-1} \cdot 2\nu_i \nu_k$, $\nu_i^{j+1}\nu_k^{j-1} + \nu_i^{j-1}\nu_k^{j+1} = \nu_i^{j-1}\nu_k^{j-1}(\nu_i^2 + \nu_k^2)$ and $2\nu_i \nu_k \leqq \nu_i^2 + \nu_k^2$.

(25) and (23) imply (21) and thus (19), by which also the second of the requirements (15) is established. In this way, the proof of the (strict) monotony of the Temple quotients is complete.

REMARK. If, moreover,

(26) $\langle F_0, N\phi_1\rangle \neq 0$,

then (see [3], (2.43), p. 224)

(27) $\lim_{j\to\infty} \mu_j = \lambda_1$.

An easy computation then yields

(28) $\lim_{j\to\infty} \tau_j(L) = \lambda_1$ for every $L \in (\mu_1, \lambda_2)$.

If λ_1 is not simple, then (26) should be fulfilled for at least one eigenelement corresponding to λ_1.

References :

[1] Goerisch, F. - Albrecht, J.: Die Monotonie der Templeschen Quotienten. ZAMM (in print).

[2] Rektorys, K.: Variational Mehods in Mathematics, Science and Engineering, 2nd Ed. Dortrecht-Boston-London, Reidel 1979.

[3] Rektorys, K. - Vospěl, Z.: On a Method of Twosided Eigenvalue Estimates for Elliptic Equations of the Form $Au - \lambda Bu = 0$. Aplikace matematiky 26(1981), 211-240.

[4] Rektorys, K.: A Proof of Monotony of the Temple Quotients in Eigenvalue Problems. Aplikace matematiky (in print).

Prof. RNDr Karel Rektorys, DrSc,
Technical University Prague,
Thákurova 7, 166 29 Praha 6.

ISNM, Vol. 69
Numerical Treatment of
Eigenvalue Problems, Vol. 3

EIGENVECTOR ESTIMATES AND APPLICATION TO SOME PROBLEMS OF STRUCTURAL ENGINEERING

Manfredi Romano

Computation of approximate eigenmodes is required in many problems of of structural engineering, so that the relevant error estimates are of great practical importance. This paper presents the complete proof of new error estimates which are based on an original technique due to G. Fichera. In the seismic analysis of structures a number of natural frequencies and associated vibration modes are computed by means of a Rayleigh - Ritz, finite element approximation. It is shown that the error estimates of this paper can be applied to the modal analysis of framed structures. To this end is presented a method to obtain an explicit representation of the Green operator that is suitable for computation by numerical procedures also in the case of very complex structures, as real buildings are.

1. INTRODUCTION

This paper deals with the problem of computing error estimates for eigenvectors in vibration problems of elastic structures. We confine ourselves to structural problems in which the Green operator can be explicitly constructed. In the first part of the paper a complete proof of new estimates, which give upper and lower bounds to the error, is given. The upper estimate was first proposed by the author in [5] as an improved version of the estimate due to G. Fichera [3]. The lower estimate was first presented in [1]. In the second part it is presented a method to obtain an explicit representation of the Green operator in a form which is suitable for computation by numerical procedures, also in the case of very complex structures.

2. BASIC ERROR REPRESENTATION IN AN ABSTRACT SETTING

Since we have in mind application to vibration problems, we shall suppose that the Green operator T of the structure is a positive compact operator on the complex Hilbert space H. Moreover to simplify the notations we shall suppose that all the eigenvalues are simple. All the arguments that follow can be easily extended to the more general case of hermitian compact operators with multiple positive and negative eigenvalues.

We shall denote by $\{\mu_h\}$ the decreasing sequence of eigenvalues of T and by $\{u_h\}$ an orthonormal sequence of associated eigenvectors. We denote by $(\cdot,\cdot)$ the inner product, by $\| \ \|$ the norm and by S_H the unit sphere in H. Let us consider the projectors $P_h = u_h \otimes u_h$ ($\otimes$ = tensor product on HxH) and the orthogonal projector P_o on ker T. The operator T has the following spectral decomposition

$$T = \Sigma \mu_h P_h \tag{2.1}$$

and every $v \in H$ has the representation

$$v = \Sigma P_h v + P_o v \tag{2.2}$$

Given $v \in S_H$ we define the error relative to the approximation of a subspace U of H by means of v, as the distance of v from U. Hence if we denote by P the projector on U and by $Q = I - P$ its orthogonal complement, the error is given by $\|Qv\|$ .We shall consider the approximation of the eigenspace associated to the eigenvalue μ_k, so that $U = sp(u_k)$, $P = P_k$ and

$$Qv = v - Pv = \sum_{h \neq k} P_h v + P_o v \tag{2.3}$$

Let us denote by Σ the spectrum of T and by Γ the resolvent set. Let Λ be the set $\{\mu_k\}$. Given $v \in S_H$, for every $\mu \in \Gamma \cup \Lambda$, we set

$$Tv - \mu v = f \tag{2.4}$$

From (2.1) we have

$$P_o Tv = TP_o v = 0, \qquad P_h Tv = TP_h v = \mu_h P_h v$$

so that, substituting (2.2) in (2.4), it follows

$$P_o v = -\frac{P_o f}{\mu}, \qquad P_h v = \frac{P_h f}{\mu_h - \mu} \quad h \neq k \tag{2.5}$$

By means of (2.5) and (2.3), we get

$$Qv = \sum_{h \neq k} \frac{P_h f}{\mu_h - \mu} - \frac{P_o f}{\mu} \tag{2.6}$$

hence

$$\| Qv \|^2 = \sum_{h \neq k} \frac{\| P_h f \|^2}{(\mu_h - \mu)^2} + \frac{\| P_o f \|^2}{\mu^2} \tag{2.7}$$

that is the representation we wanted to obtain.

3. THE UPPER ESTIMATE

We have

$$\sum_{h \neq k} \| P_h f \|^2 + \| P_o f \|^2 = \| Qf \|^2 \leq \| f \|^2 \tag{3.1}$$

$$\| f \|^2 = (Tv, Tv) - 2\mu\,(Tv, v) + \mu^2 \tag{3.2}$$

If we define

$$\beta = ||Tv|| \qquad\qquad \alpha = (Tv,v) \tag{3.3}$$

the relation (3.2) can be written in the form

$$||f||^2 = \beta^2 - 2\mu\alpha + \mu^2 = \beta^2 - \alpha^2 + (\mu - \alpha)^2 \tag{3.4}$$

Making use of (3.1) and (3.4), from the representation (2.7), we get the following upper estimate for $||Qv||^2$

$$||Qv||^2 \leq \frac{\beta^2 - \alpha^2 + (\mu - \alpha)^2}{\min\limits_{\gamma \in \Sigma - \Lambda} (\gamma - \mu)^2} = g(\mu) \tag{3.5}$$

Let us notice that by the Cauchy inequality we have $\beta^2 - \alpha^2 \geq 0$.

To obtain an optimal upper estimate we seek the minimum of $g(\mu)$. We shall prove that a necessary condition for the estimate (3.5) to be non trivial, is that

$$\mu \,,\, \alpha \in \,]\, \mu_{k+1} \,,\, \mu_k \,[\tag{3.6}$$

Let us first suppose that μ doesn't satisfy the condition (3.6). From (2.1) and (2.2) we have

$$\beta^2 = \sum_h \mu_h^2 \, ||P_h v||^2 , \qquad\qquad \alpha = \sum_n \mu_h \, ||P_h v||^2 \tag{3.7}$$

Substituting (3.7) in (3.4) we see that the numerator of $g(\mu)$ can be written in the form

$$\sum_h (\mu_h^2 - 2\mu\mu_h) \, ||P_h v||^2 + \mu^2 \tag{3.8}$$

which is a linear functional in the variables $||P_h v||^2$ on the closed convex polyhedron defined by the inequalities

$$|| P_h v||^2 \geq 0, \qquad \sum_h ||P_h v||^2 \leq 1 \tag{3.9}$$

The extreme points of this convex set are given by the intersections of the linear variety defined by the relation $\sum_h ||P_h v||^2 = 1$ with the coordinate axes, and by the origin. On the i-th coordinate axis it is $||P_i v||^2 = 1$ and $||P_j v||^2 = 0$ for every $j \neq i$, hence the functional (3.8) has the value $(\mu_i - \mu)^2$. In the origin it is $||P_i v||^2 = 0$ for every i, and the functional (3.8) has the value μ^2. Since μ doesn't satisfy (3.6) the values $(\mu_i - \mu)^2$ for $i \neq k$ and μ^2 are never less than the denominator of $g(\mu)$, what implies that

$$\inf_\mu \ g(\mu) \geq 1 \tag{3.10}$$

If μ satisfies the condition (3.6) and α doesn't it is easy to see from (3.5) that again $(\mu - \alpha)^2$ is not less than the denominator of $g(\mu)$, what gives $g(\mu) \geq 1$, and the proof is complete.

We have now to seek the minimum of $g(\mu)$ under the condition (3.6). Let us denote by γ_1 and γ_2 respectively a lower bound to μ_k and an upper bound to μ_{k+1} and by $\delta = (\gamma_1 + \gamma_2)$ the middle point of the interval (γ_1, γ_2).

To obtain computable upper estimates we must substitute the condition (3.6) by the stronger one

$$\mu, \alpha \notin \,]\gamma_2 , \gamma_1[\tag{3.11}$$

Let us define

$$g_k(\mu) = \frac{\beta^2 - \alpha^2 + (\mu - \alpha)^2}{(\mu - \gamma_k)} \qquad k = 1,2 \tag{3.12}$$

We have $g(\mu) = g_2(\mu)$ if $\gamma_2 < \alpha \leq \delta$ and $g(\mu) = g_1(\mu)$ if $\delta \leq \alpha < \gamma_1$. Hence we shall consider the following two cases.

Case 1 : $\gamma_2 < \alpha \leq \delta$

In this case the function $g(\mu) = g_2(\mu)$ has the minimum in the point

$$\mu = \mu_{02} = \alpha + \frac{\beta^2 - \alpha^2}{\alpha - \gamma_2}$$

If $\mu_{02} \leq \delta$, i.e. if $\beta^2 - \alpha^2 \leq (\delta - \alpha)(\alpha - \gamma_2)$, the minimum of $g(\mu)$ is

$$g(\mu_{02}) = \frac{\beta^2 - \alpha^2}{(\alpha - \gamma_2) + \beta^2 - \alpha^2} < 1 \tag{3.13}$$

If $\mu_{02} \geq \delta$, the minimum of $g(\mu)$ is given by

$$g(\delta) = \frac{\beta^2 - \alpha^2 + (\delta - \alpha)^2}{(\gamma_1 - \gamma_2)/2} \tag{3.14}$$

Case 2 : $\delta \leq \alpha < \gamma_1$

In this case the function $g(\mu) = g_1(\mu)$ has the minimum in the point

$$\mu = \mu_{01} = \alpha - \frac{\beta^2 - \alpha^2}{\gamma_1 - \alpha}$$

If $\mu_{01} \geq \delta$, i.e. if $\beta^2 - \alpha^2 \geq (\alpha - \phi)(\gamma_1 - \alpha)$, the minimum of $g(\mu)$ is

$$g(\mu_{01}) = \frac{\beta^2 - \alpha^2}{(\gamma_1 - \alpha)^2 + \beta^2 - \alpha^2} < 1 \tag{3.15}$$

If $\mu_{01} \leq \delta$, the minimum of $g(\mu)$ is given by (3.14).

The optimal upper estimate is given by (3.13), (3.14) and (3.15) and requires the knowledge of α and β , defined by (3.3), of the lower bound γ_1 and of the upper bound γ_2. These bounds must be good enough to separate the eigenvalue μ_k from the others.

4. THE LOWER ESTIMATE

Let us set in (2.4) $\mu = \mu_k$. Then we have

$$Pf = P(Tv - \mu_k v) = 0, \qquad Qf = f$$

It follows that

$$\sum_{h \neq k} \|P_k f\|^2 + \| P_o f \|^2 = \| Qf \|^2 = \|f\|^2 \tag{4.1}$$

From (2.7) and (4.1) we get the non trivial lower estimate

$$\| Qv \|^2 \quad \frac{\beta^2 - \alpha^2 + (\mu_k - \alpha)^2}{\max \{(\mu_1 - \mu_k)^2, \mu_k^2\}} \tag{4.2}$$

Let us now suppose that we have at our disposal a lower bound μ_k' and an upper bound μ_k'' that separate the eigenvalue μ_k from the others. Then we obtain non trivial computable lower estimates from (4.2) in the following way. We substitute the denominator of (4.2) with the greater number given by

$$\max \ \{(\mu_k'' - \mu_k'), \mu_k''\}$$

and the quantity $(\mu_k - \alpha)^2$ at the numerator of (4.2) with smaller numbers according to the scheme

$$(\mu_k' - \alpha)^2 \qquad \text{if} \quad \alpha \leq \mu_k'$$

$$(\mu_k'' - \alpha)^2 \qquad \text{if} \quad \mu_k'' \leq \alpha$$

$$0 \qquad \text{if} \quad \mu_k' \leq \alpha \leq \mu_k''$$

5. A CLASS OF STRUCTURAL PROBLEMS

Approximate vibration modes of elastic structures are usually computed by a Rayleigh - Ritz, finite element procedure. We consider here structures whose elements can described by a beam model. This class is of great interest in engineering practice.

We want to show how to construct an explicit representation of the Green function for the relevant vibration problem. Let ω be the angular frequency. We set $\lambda = \omega^2$, $\mu = \lambda^{-1}$ and introduce the following notations

n	number of beam elements
$v(x)$	n - vector valued displacement function of the beams
s	r - vector of the Lagrangean parameters in the finite element approximation
V	space of admissible displacement functions
k_i	elastic compliance of the cross sections of the i-th beam
m_i	distributed mass on the axis of the i-th beam
ℓ_i	lenght of the axis of the i-th beam
M	concentrated mass matrix, such that λMs is the vector of the associated inertia forces

The eigenvalue problem has the following variational formulation

$$B\,(v\;,\;v^*) = \lambda\;(\;Ms\cdot s^* + m\;(v,v^*)\;) \qquad (v,v^*) \in V \times R^r \qquad (5.1)$$

where $B\,(v\;,\;v^*)$ is the bilinear form of the internal virtual work.

If we define

$$q_i = \lambda\; m_i\; v_i, \qquad\qquad f_i = \lambda\;(Ms)_i \qquad (5.2)$$

and substitute in (5.1), we get the variational formulation of the associated static problem in the form

$$B\ (v\ ,\ v^*) = f \cdot s + (q, v^*) \tag{5.3}$$

The problem (5.3) can be split in the following two

$$B\ (v'\ ,\ v^*) = (q, v^*) \tag{5.4}$$

$$B\ (v''\ ,\ v^*) = f \cdot s \tag{5.5}$$

Hence

$$v = v' + v'', \qquad s = s' + s'' \tag{5.6}$$

Let us now introduce on the beam axes the adimensional abscissa $x = z_i/\ell_i$, so that integrations will be performed on the interval (0,1).

From the differential formulation of the problem (5.4) we know that v_i' must be of the form

$$v_i'\ (x) = C_i s' \cdot t(x) \ell_i + \ell_i^3\ k_i \int_0^1 \bar{g}(x,y)\ q(y)\ dy \tag{5.7}$$

where C_i is the Boolean matrix that maps s' in the vector of nodal parameters for the i-th beam and t(x) is the vector of displacement functions associated to the nodal parameters. By C_i' we shall denote the traspose of C_i. The function $\bar{g}(x,y)$ is the Green kernel for the clamped beam. Solving (5.4) for s' we find that

$$s' = E\ C_j'\ \ell_j^2 \int_0^1 t(y)\ q_i(y)\ dy \tag{5.8}$$

where E is the compliance matrix of the finite element model of the structure.

From the differential formulation of the problem (5.5) we know that v_i''

must be of the form

$$v''_i(x) = C_i s'' t(x) \ell_i \tag{5.9}$$

and solving (5.5) for s" we get

$$s'' = E\, f \tag{5.10}$$

We may now substitute (5.8) in (5.7) and (5.10) in (5.9) to get from (5.6) and (5.2) an equivalent eigenvalue problem for an integral operator. To this aim let us consider the two sets $A_1 = \underset{i}{X}(0,1)$ and $A_2 = \{a\}$ where a is a real number. On A_1 we have the Lebesgue measure dx=dy, on {a} the Dirac measure da. If we define $A = A_1 \cup A_2$, the eigenvalue problem can be put in form

$$H\, z = \int_A h(x,y)\, z(y)\, dy = \lambda z \tag{5.11}$$

where z = (v;s) and the matrix kernel h(x,y) is defined as follows

$$\begin{aligned}
h_{11}(x,y) &= \{\ell_i\, g_{ij}(x,y)\, m_j \ell_j^2\} && \text{on } A_1 \times A_1 \\
h_{12}(x,a) &= \{M\, E\, \ell_i C'_i\, t(x)\} && \text{on } A_1 \times A_2 \\
h_{21}(a,y) &= \{E\, C'_i\, t(y)\, m_i \ell_i^2\} && \text{on } A_2 \times A_1 \\
h_{22}(a,a) &= \{E\, M\} && \text{on } A_2 \times A_2
\end{aligned} \tag{5.12}$$

and

$$g_{ij}(x,y) = C_i\, E\, C'_j\, t(y) \cdot t(x) + \delta_{ij}\, \ell_j\, k_j\, \bar{g}(x,y) \tag{5.13}$$

In (5.12) and (5.13) sum on repeated j is implicit and δ_{ij} is the unit matrix.

Let a normalized r - vector c be an approximate eigenvector computed by the finite element method and define

$$w_i(x) = C_i c\ t(x)\ell_i, \qquad z = (w;c) \tag{5.14}$$

We want to compute

$$\beta^2 = (Hz\ ,\ Hz)/(z\ ,\ z)$$

$$\alpha = (Hz\ ,\ z)/(z\ ,\ z)$$

By means of (5.14), (5.11), (5.12) and (5.13) we get

$$(z\ ,\ z) = (P_2 + I)\ c \cdot c$$

$$(Hz\ ,\ z) = (P_2EP_1 + H_1 + MEP_2 + EP_1 + EM)\ c \cdot c$$

$$(Hz\ ,\ Hz) = (P_1EP_2EP_1 + H_3 + 2H_2EP_1 + MEP_2EM + 2ME(P_2EP_1 + H_1))\ c \cdot c$$

where I is the unit matrix and the matrices $E, P_1, P_2, H_1, H_2, H_3$ are defined by the following relations

$$E = C_i'\ K\ C_i$$

$$P_1 = C_i'\ m_i\ \ell_i^3\ L\ C_i$$

$$P_2 = C_i'\ \ell_i^3\ L\ C_i$$

$$H_1 = C_i'\ m_i\ \ell_i^7\ k_i\ G_1\ C_i$$

$$H_2 = C_i' m_i \ell_i k_i G_2 C_i$$

$$H_3 = C_i' m_i^2 \ell_i^{11} k_i^2 G_3 C_i$$

In the above definitions the sum on the repeated indexes i is implicit. The matrices K, L, G_1, G_2, G_3 can be easily computed from the following explicit relations

$$K = \int_0^1 D^2 t(x) \otimes D^2 t(x)\, dx$$

$$L = \int_0^1 t(x) \otimes t(x)\, dx$$

$$G_1 = \int_0^1 \int_0^1 \bar{g}(x,y)\, t(x) \otimes t(y)\, dy\, dx$$

$$G_2 = \int_0^1 \left(\int_0^1 \bar{g}(x,y)\, t(y)\, dy \otimes t(x) \right) dx$$

$$G_3 = \int_0^1 \left(\int_0^1 \bar{g}(x,y)\, t(y)\, dy \otimes \int_0^1 \bar{g}(x,y)\, t(y)\, dy \right) dx$$

The matrices E and P_1 are respectively the compliance and the mass matrix of the finite element model of the structure and hence must be known in order to compute the approximating vector c. All the matrices $E, P_1, P_2, H_1, H_2, H_3$ can be easily computed by techniques that are typical of computer programming, without performing the matrix products that appear in their definitions. Indeed the matrices C_i are really never created, with great saving of time and memory requirements.

REFERENCES

[1] Esposito, A. & Romano, M. : Error Estimates for Eigenvector Approximation in the Theory of Elastic Vibrations. Rendiconti di Matematica (1) Vol. 11, Serie VI (1978), 39-56.

[2] Fichera, G. : Abstract and Numerical Aspects of Eigenvalue Theory. Dept. of Math. The University of Alberta, Edmonton, Canada, (1973), 1-119.

[3] Fichera, G. : Approximation of the Eigenvectors of a Positive Compact Operator and Estimate of the Error. Annali di Matematica Pura ed Applicata, Serie IV, Tomo CVIII, (1976), 367-377.

[4] Romano, M. : Upper and Lower Bounds to the Eigenfrequencies of Elastic Frames. Meccanica, N° 3, Vol. X, (1975), 203-209.

[5] Romano, M. : An Optimal Error Estimate for Eigenvectors in Buckling and Vibration of Elastic Structures. Atti Ist. Scienza delle Costruz., N° 275, Fac. di Ing., Napoli, (1977), 1-12.

[6] Weinberger, H.F. : Variational Methods for Eigenvalue Approximation, Regional Conference Series in Applied Mathematics, 15, SIAM, Phyladelphia, 1974.

M. Romano
Ist. di Scienza delle Costruzioni
Facoltà di Ingegneria
Viale Andrea Doria 6
95125 Catania
ITALY

ISNM, Vol. 69
Numerical Treatment of
Eigenvalue Problems, Vol. 3

EINE VARIANTE DES LANCZOS-VERFAHRENS

Hans-Rudolf Schwarz

The computation of eigenfrequencies and eigenmodes of technical systems by the method of finite elements leads to generalized eigenvalue problems $Ax = \lambda Bx$ with symmetric and sparse matrices A and B of high order n. If the classical Lanczos algorithm is applied to a sequence of inverse and shifted eigenvalue problems with appropriate chosen shifts, the resulting method is indeed very efficient. Each Lanczos run should yield only a certain group of eigenvalues and eigenvectors. In case of small group sizes reorthogonalization of the iterated basis vectors is unnecessary due to a small number of Lanczos steps required. The proposed version combines some techniques used in the bisection method in order to increase the reliability and efficiency of the process. Some examples show the superiority of the algorithm compared with other methods.

1. Problemstellung und Beispiele

Zur Motivation der nachfolgenden Betrachtungen sollen einige technische Fragestellungen dienen. Um das dynamische Verhalten eines schwingungsfähigen Systems zu erfassen, sind sowohl seine Eigenschwingsungsfrequenzen als auch seine Eigenschwingungsformen zu berechnen, die etwa die Grundlage für eine Modalanalyse bilden. Solche Untersuchungen sind bekanntlich notwendig, um entweder nur störende oder aber zur Zerstörung des Systems führende Resonanerscheinungen ausschliessen zu können.

Beispiel 1. Um das unangenehme Dröhnen im Innern eines Autos durch geeignete Massnahmen zu reduzieren, interessiert sich die Autoindustrie für die akustischen Eigenfrequenzen und die zugehörigen Stehwellen im Autoinnern. Nach Separation der räumlichen Variablen ist für einen zweidimensionalen Längsschnitt die

Helmholtz-Gleichung $\Delta u + \lambda u = 0$ unter Neumannschen Randbedingungen $\frac{\partial u}{\partial n} = 0$ zu behandeln. Tut man dies mit der Methode der finiten Elemente auf Grund der in Figur 1 dargestellten Dreieckeinteilung, verwendet in in jedem Dreieckelement kubische Ansätze mit neun Freiheitsgraden [8], so ergeben sich bei 305 Knotenpunkten $n = 915$ Knotenvariable.

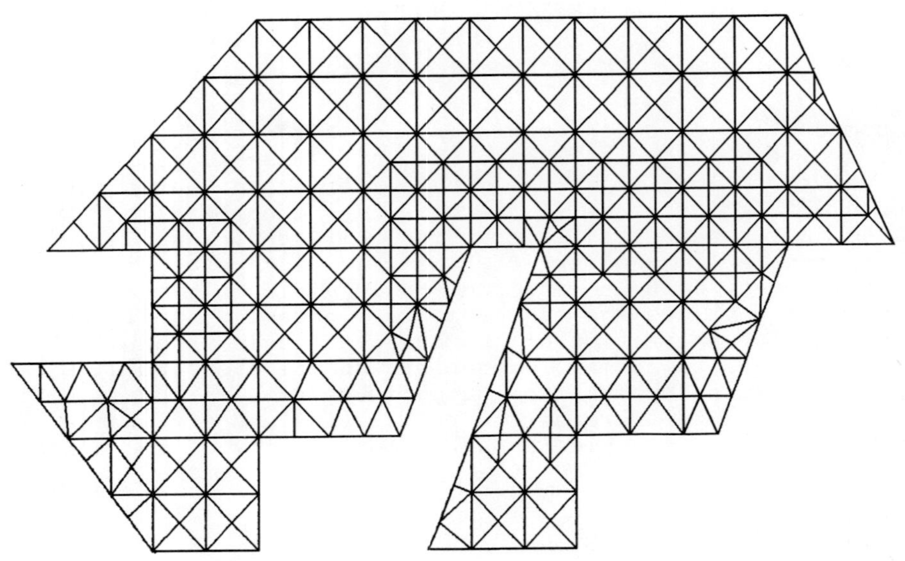

Figur 1 Autolängsschnitt mit Dreieckeinteilung

<u>Beispiel 2</u>. In Erdbebengebieten ist es wichtig, die Eigenfrequenzen von Hochbauten vor deren Bau zu analysieren. Wir betrachten die in Figur 2 dargestellte, einfache und durchschaubare Eisenbalkenkonstruktion, wie man sie beim modernen Bürohausbau antrifft. Die Konstruktion setzt sich in natürlicher Weise aus Balkenelementen zusammen, die auf Biegung in zwei Richtungen, Torsion und Zug beansprucht werden. In jedem Knoten sind sechs Freiheitsgrade zu berücksichtigen, sodass unter der Annahme von 24 im Fundament eingespannten Fusspunkten bei 144 freien Knotenpunkten $n = 864$ Knotenvariable resultieren.

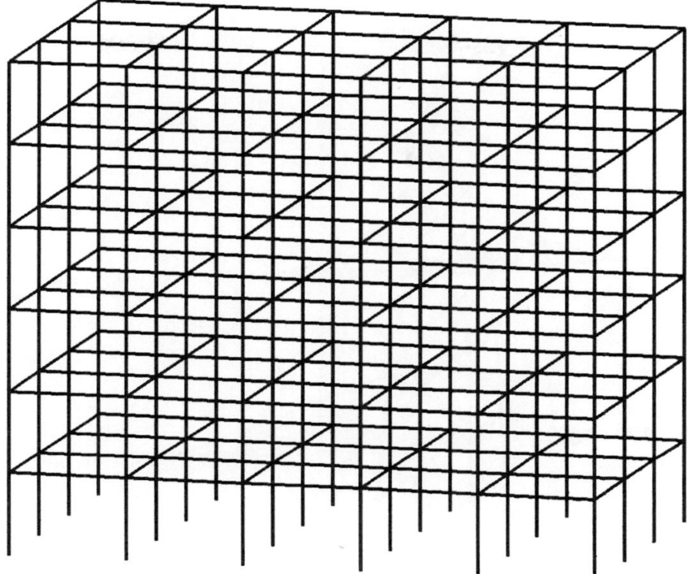

Figur 2 Einfache Balkenkonstruktion, Bürohaus

Beispiel 3. Auf Grund derselben Problematik interessieren die Eigenfrequenzen und Eigenschwingungsformen eines in den Bergen stehenden Hochspannungsmastes. Die Figur 3 zeigt seinen Aufbau, wobei die drei verschiedenen Strichdicken die unterschiedlichen Querschnittdimensionierungen der betreffenden Balkenelemente andeutet. Die Höhe des Mastes vom tiefsten Fusspunkt bis zur Spitze beträgt rund 47 Meter. Die 167 Knotenpunkte, in denen die Balkenelemente starr miteinander verbunden sind, sind durch ausgefüllte Kreise hervorgehoben. Da die vier Fusspunkte realistischerweise als im Fels eingespannt angenommen sind, resultieren bei 163 freien Knotenpunkten insgesamt n = 978 Knotenvariable.

2. Mathematische Aufgabenstellung

Schwingungsaufgaben der beschriebenen Art, welche unter der Annahme von kleinen Auslenkungen auf Grund der linearen Theorie mit der Methode der finiten Elemente unter Verwendung von konsistenten Massenmatritzen behandelt werden, führen zu allge-

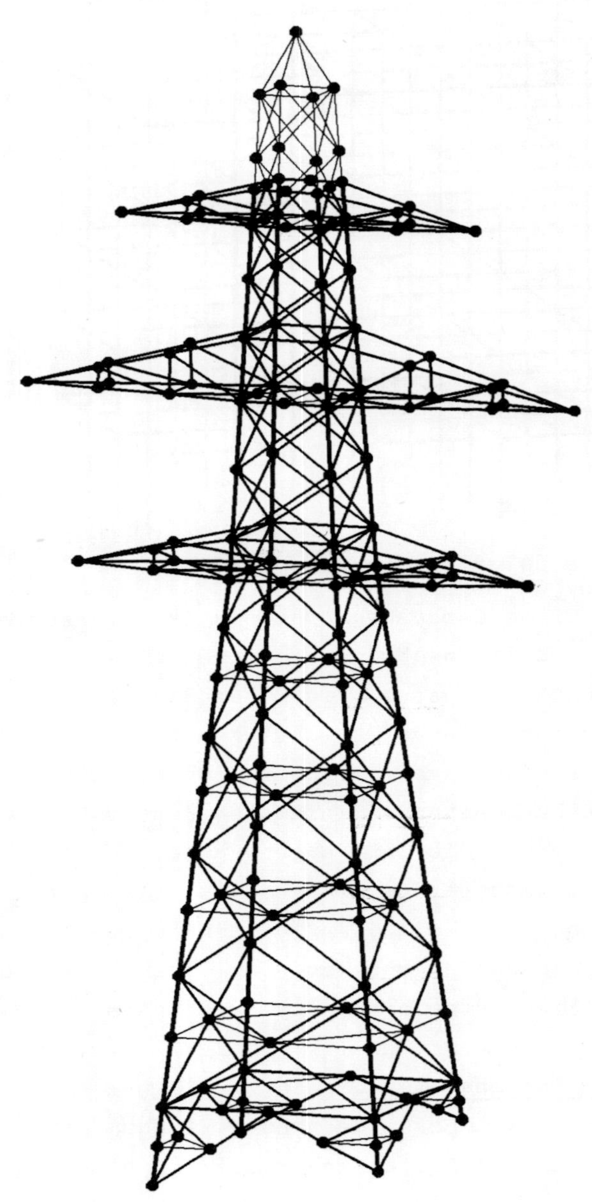

Figur 3 Hochspannungsmast

meinen Eigenwertaufgaben

$$A \underline{x} = \lambda B \underline{x} . \tag{1}$$

Da die Zahl der Knotenvariablen entweder infolge der getroffenen feinen Einteilung in Elemente oder auf Grund der gegebenen Konstruktion des Systems hoch ist und in praktischen Anwendungen mehrere hundert bis mehrere tausend beträgt, ist die Ordnung n der Matrizen A und B entsprechend gross. Die Matrizen A und B sind symmetrisch und in der Regel positiv definit. Als unmittelbare Folge der Methode der finiten Elemente sind sie schwach besetzt, wobei sie die gleiche Besetzungsstruktur aufweisen, d.h. ihre von Null verschiedenen Matrixelemente stehen an den gleichen Stellen.

In bestimmten Anwendungen werden die p kleinsten Eigenwerte $\lambda_1 \leq \lambda_2 \leq \ldots \leq \lambda_p$ der Eigenwertaufgabe (1) gesucht, die den kleinsten Kreisfrequenzen $\omega_1 \leq \omega_2 \leq \ldots \leq \omega_p$ des Systems mit $\omega_i^2 = \lambda_i$ entsprechen, sowie die zugehörigen Eigenvektoren $\underline{x}_1$, $\underline{x}_2, \ldots \underline{x}_p$, welche die entsprechenden Schwingungsformen beschreiben. Bei Resonanzproblemen sind alle Eigenfrequenzen ω_i zu bestimmen, welche in einem vorgegebenen Intervall [a,b] liegen. In diesem Fall sind die p Eigenwerte $\lambda_r \leq \lambda_{r+1} \leq \ldots \leq \lambda_{r+p-1}$ mit den zugehörigen Eigenvektoren $\underline{x}_r, \underline{x}_{r+1}, \ldots , \underline{x}_{r+p-1}$ gesucht, wobei die Indizes durch das Intervall [a,b] festgelegt sind. In beiden Fällen ist die Anzahl p der gesuchten Eigenwerte im Vergleich zur Ordnung n der Matrizen sehr klein.

Um die formulierte Aufgabe numerisch möglichst effizient zu lösen, ist die schwache Besetzung der Matrizen A und B oder zumindest eine bestimmte Besetzungsstruktur auszunützen. Es ist klar, dass die Besetzungsstruktur der Matrizen wesentlich von der Numerierung der Knotenvariablen abhängt. So ist die Bandstruktur oder in Verfeinerung die Hüllenstruktur sowohl für den Speicherbedarf als auch für den Rechenaufwand massgebend. Im folgenden konzentrieren wir uns auf die Hüllenstruktur, mit welcher die in den praktischen Anwendungen häufig anzutreffende stark variierende Bandbreite erfasst wird. Dazu definiert man mit

$$f_i(A) := \min \{ j \mid a_{ij} \neq 0, j \leq i \}, \quad (i=1,2,\ldots,n) \tag{2}$$

die Kolonnenindizes der ersten von Null verschiedenen Matrixelemente a_{ij} jeder Zeile. Dann bezeichnet man als die Hülle oder Enveloppe einer Matrix A

$$\mathrm{Env}(A) := \{(i,j) \mid f_i(A) \leq j \leq i,\ 1 \leq i \leq n\} \tag{3}$$

die Menge der Indexpaare (i,j), für welche die zugehörigen Matrixelemente a_{ij} vom ersten von Null verschiedenen bis zum Diagonalelement in jeder Zeile liegen. Die Anzahl dieser Indexpaare

$$P := |\mathrm{Env}(A)| \tag{4}$$

nennt man das Profil der Matrix A. Das Profil bestimmt im wesentlichen den Speicherbedarf für eine symmetrische, positiv definite Matrix, um etwa ihre Cholesky-Zerlegung zu berechnen und zu speichern [8].

Werden die Knotenpunkte des Autolängsschnittes von Beispiel 1 mit Hilfe des Algorithmus von Cuthill-McKee [8,9] optimal numeriert, um ein minimales Profil zu erzielen, resultiert eine ausgesprochene Bandstruktur mit nicht allzu stark veränderlicher Bandbreite. Das Profil beträgt P = 34320, während die Anzahl der von Null verschiedenen Matrixelemente der unteren Hälfte N = 9291 ist.

Falls die Knotenpunkte im Beispiel 2 systematisch in aufsteigenden Ebenen durchnumeriert werden, tritt der regelmässige Aufbau der dreidimensionalen Konstruktion in der Besetzungsstruktur augenfällig in Erscheinung. Die Matrizen besitzen eine klare Bandstruktur mit einer (einseitigen) Bandbreite m = 149, wobei jedoch das Innere des Bandes sehr spärlich besetzt ist. Diese Tatsache äussert sich auch im Profil P = 110772, während die Zahl der von Null verschiedenen Matrixelemente der unteren Hälfte nur N = 15552 beträgt.

Werden die Knotenpunkte des Hochspannungsmastes von Beispiel 3 sukzessive in Ebenen von oben nach unten durchnumeriert, erhalten die Matrizen A und B eine Besetzungsstruktur gemäss Figur 4, wobei jedes Quadrat einer Untermatrix der Ordnung sechs entspricht. Die drei Mastausleger sind übrigens deutlich zu erkennen, und in diesem Beispiel liegt eindeutig eine Hüllenstruktur vor. Die Anzahl der von Null verschiedenen Matrixelemente der unteren Hälfte ist N = 20955, während das Profil P = 48855 nur gut doppelt so gross ist.

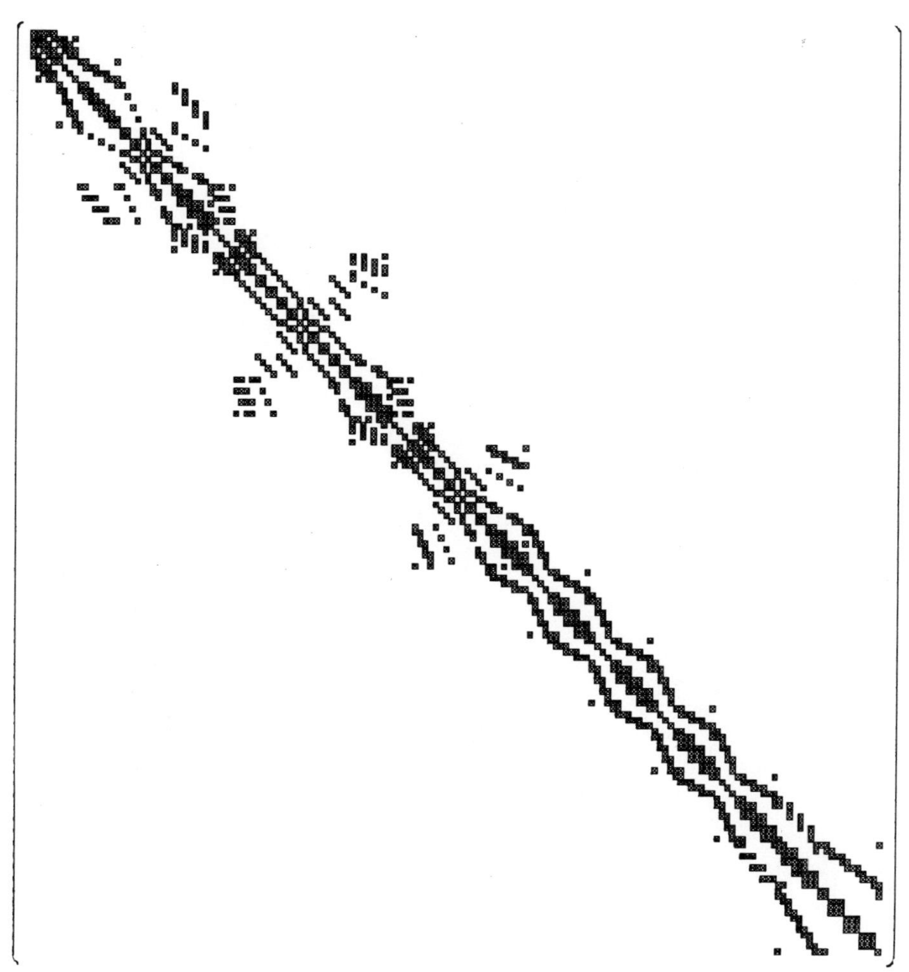

Figur 4 Besetzungsstruktur für Hochspannungsmast

3. Der Lanczos-Algorithmus

Im Jahr 1950 hat Lanczos [3] ein Verfahren vorgeschlagen zur Berechnung der Eigenwerte λ_k einer symmetrischen und schwach besetzten Matrix A grosser Ordnung. Ausgehend von einem Startvektor $\underline{f}_1 \neq \underline{0}$ bildete er die iterierten Vektoren

$$\underline{f}_2 = A\,\underline{f}_1,\ \underline{f}_3 = A\,\underline{f}_2,\ \ldots,\ \underline{f}_m = A\,\underline{f}_{m-1}, \tag{5}$$

d.h. die Krylov-Folge

$$K^{(m)}(\underline{f}_1) := \{\underline{f}_1,\ A\,\underline{f}_1,\ A^2\underline{f}_1,\ \ldots,\ A^{m-1}\underline{f}_1\}. \tag{6}$$

Unter der Voraussetzung der linearen Unabhängigkeit lässt sich in dem von der Krylov-Folge aufgespannten Unterraum eine orthonormierte Basis $\underline{q}_1, \underline{q}_2, \ldots, \underline{q}_m$ konstruieren. Sie lässt sich sukzessive so bilden, dass im allgemeinen k-ten Schritt lediglich der Vektor $\underline{u}_k := A\,\underline{q}_{k-1}$ gegenüber den beiden vorhergehenden Basisvektoren $\underline{q}_{k-1}$ und $\underline{q}_{k-2}$ zu orthogonalisieren ist. Denn die Folge der Basisvektoren erfüllt die Rekursionsformel ($\underline{q}_o = \underline{0}$)

$$A\,\underline{q}_i = \beta_{i-1}\underline{q}_{i-1} + \alpha_i\underline{q}_i + \beta_i\underline{q}_{i+1}, \quad (i=1,2,\ldots,m-1). \tag{7}$$

Fasst man die m orthonormierten Vektoren $\underline{q}_k$ zur Matrix

$$Q_m := (\underline{q}_1, \underline{q}_2, \ldots, \underline{q}_m) \in \mathbb{R}^{n\times m} \tag{8}$$

zusammen, dann gilt wegen der Orthonormiertheit und wegen (7)

$$T_m := Q_m^T\,A\,Q_m = \begin{bmatrix} \alpha_1 & \beta_1 & & & & \\ \beta_1 & \alpha_2 & \beta_2 & & & \\ & \beta_2 & \alpha_3 & \beta_3 & & \\ & & \cdot & \cdot & \cdot & \\ & & & \cdot & \cdot & \cdot \\ & & & \beta_{m-2} & \alpha_{m-1} & \beta_{m-1} \\ & & & & \beta_{m-1} & \alpha_m \end{bmatrix} \in \mathbb{R}^{m\times m}. \tag{9}$$

Die tridiagonale Matrix T_m ist wegen $\beta_i \neq 0$ $(i=1,2,\ldots,m-1)$ irreduzibel. T_m ist genau diejenige Matrix, welche bei einem Ritz-Schritt mit den m orthonormierten Vektoren $\underline{q}_1, \underline{q}_2, \ldots, \underline{q}_m$ resultiert, um auf diese Weise Eigenwerte und Eigenvektoren der gegebenen Matrix zu approximieren. Lösen wir das Eigenwertproblem

$$T_m \underline{s} = \Theta \underline{s}, \tag{10}$$

so stellen die m Eigenwerte Θ_k, welche als Ritzwerte bezeichnet werden, in einem bestimmten Sinn die bestmöglichen Approximationen für gewisse Eigenwerte λ_i von A dar. Desgleichen sind die Ritzvektoren

$$\underline{y}_k := Q_m \underline{s}_k, \quad (k=1,2,\ldots,m), \tag{11}$$

welche mit den entsprechenden Eigenvektoren $\underline{s}_k$ von (10) gebildet werden, bestmögliche Näherungen der zu λ_i gehörenden Eigenvektoren $\underline{x}_i$ von A [7].

Das klassische Lanczos-Verfahren besitzt für das spezielle Eigenwertproblem A $\underline{x} = \lambda \underline{x}$ die einfache algorithmische Beschreibung:

<u>Start</u>: Wähle $\underline{r}_o \neq \underline{0}$; setze $\underline{q}_o = \underline{0}$, $\beta_o = \| \underline{r}_o \|$.

<u>Iteration</u>: (für j = 1,2,3,...)

1. $\underline{q}_j = \underline{r}_{j-1} / \beta_{j-1}$ (Normierung)
2. $\underline{u}_j = A \underline{q}_j$
3. $\underline{r}_j = \underline{u}_j - \beta_{j-1} \underline{q}_{j-1}$ (Orth. bez. $\underline{q}_{j-1}$) (12)
4. $\alpha_j = \underline{q}_j^T \underline{r}_j$
5. $\underline{r}_j := \underline{r}_j - \alpha_j \underline{q}_j$ (Orth. bez. $\underline{q}_j$)
6. $\beta_j = \| \underline{r}_j \|$
7. ev. Ritzwerte/-vektoren aus T_j, Q_j ;
   Test (Konvergenz Ritzwerte/-vektoren)

Die theoretischen Eigenschaften des Algorithmus gehen infolge der endlichen Genauigkeit der numerischen Durchführung verloren. So stellt man fest, dass die paarweise Orthogonalität der Vektoren $\underline{q}_j$ früher oder später nicht mehr erfüllt ist, und dass die Basisvektoren sogar linear abhängig werden. Deshalb galt der Lanczos-Algorithmus (12) während längerer Zeit als instabil und somit als unbrauchbar. Diese Ansicht änderte sich, nachdem in den vergangenen Jahren die wirklichen Gründe erforscht und folgende Tatsachen aufgezeigt worden sind [4,5,6,7]:

Die Orthogonalität der Vektoren $\underline{q}_j$ muss verloren gehen, sobald ein Ritzpaar (Θ_k, $\underline{y}_k$) gegen ein Eigenpaar (λ_i, $\underline{x}_i$) von A konvergiert ist. Bei Fortsetzung des Verfahrens vergisst der Prozess, dass er bereits ein Eigenpaar berechnet hat und beginnt es von neuem zu bestimmen. Es entstehen so mehrfache Kopien, wobei aber die fast identischen Ritzwerte und Ritzvektoren dennoch gute Näherungen sind. Schliesslich hat der Lanczos-Algorithmus die Tendenz, für Eigenwerte an den Enden des Spektrums rasch gute Näherungswerte zu liefern, was mit der Verteilung der Eigenwerte begründet werden kann.

Diese Erkenntnisse haben den Lanczos-Algorithmus wieder aufleben lassen, indem er durch entsprechende, teilweise aufwendige Ergänzungen zu einem brauchbaren und effizienten Verfahren modifiziert wurde. Als wichtigste und am meisten angewandte Massnahme ist die vollständige Nachorthogonalisierung der Basisvektoren $\underline{q}_j$ zu allen Vorgängern zu erwähnen. Zur Reduktion des Rechenaufwandes ist auch eine selektive Nachorthogonalisierung auf Grund von Zwischenresultaten vorgeschlagen worden [7].

Schliesslich ist zur Motivation der weiteren Überlegungen noch auf eine weitere Tatsache hinzuweisen. Werden die p kleinsten Eigenwerte einer Matrix gesucht, so stellt man bei bestimmten Eigenwertverteilungen fest, dass die Ritzwerte Θ_k zuerst gegen die grössten Eigenwerte λ_i konvergieren. Zur Bestimmung der gewünschten Eigenwerte mit Hilfe des Lanczos-Algorithmus muss zum inversen Eigenwertproblem übergegangen werden.

Im folgenden soll eine Variante vorgestellt werden, bei welcher bestimmte Eigenschaften des klassischen Lanczos-Verfahrens vorteilhaft ausgenützt werden, während sich bekannte Schwächen gar nicht bemerkbar machen sollen.

4. Lanczos-Algorithmus für inverse, spektralverschobene Eigenwertprobleme

Nach einer Idee von P. Waldvogel besteht die neue Zielsetzung darin, gezielt bestimmte Gruppen von Eigenwerten aus dem Spektrum von (1) zu berechnen, wobei gleichzeitig die Zahl der dazu erforderlichen Lanczos-Schritte möglichst klein gehalten werden soll. Um dies zu erreichen, ist es naheliegend, zu inversen, spektralverschobenen Eigenwertproblemen überzugehen, wie dies auch in [1,2] vorgeschlagen wird.

Ausgehend von der gegebenen allgemeinen Eigenwertaufgabe (1) betrachten wir zunächst das um μ spektralverschobene Problem

$$(A - \mu B)\, \underline{x} = (\lambda - \mu)\, B\, \underline{x}, \tag{13}$$

wobei wir $\mu \neq \lambda_k$ $(k=1,2,\ldots,n)$ voraussetzen wollen, damit die Matrix $(A - \mu B)$ regulär ist. Mit der Cholesky-Zerlegung der positiv definiten Matrix $B = L\, L^T$ führen wir (13) in der bekannten Art formal auf das spezielle Eigenwertproblem

$$C\, \underline{y} := L^{-1}(A - \mu B)L^{-T}\underline{y} = (\lambda - \mu)\underline{y} \text{ mit } \underline{y} = L^T\, \underline{x} \tag{14}$$

mit symmetrischer Matrix C zurück. Schliesslich betrachten wir das zu (14) inverse Eigenwertproblem

$$C^{-1}\underline{y} = L^T(A - \mu B)^{-1}L\, \underline{y} = \frac{1}{\lambda-\mu}\, \underline{y} =: \rho\, \underline{y}. \tag{15}$$

Damit liegt eine spezielle Eigenwertaufgabe mit regulärer, symmetrischer Matrix C^{-1} vor. Die Eigenwerte λ_k, welche am nächsten bei μ liegen, entsprechen den betragsgrössten Eigenwerten ρ_k von C^{-1}, welche sich zudem, in der Regel sehr gut getrennt und isoliert, an den Enden des Spektrums von C^{-1} befinden. Der klassische Lanczos-Algorithmus, angewandt auf C^{-1} stellt somit ein geeignetes Instrument dar, die betragsgrössten Eigenwerte ρ_k zu bestimmen, aus denen sich die gewünschten, dem gegebenen Wert μ benachbarten Eigenwerte λ_k gemäss (15) zu

$$\lambda_k = \mu + \frac{1}{\rho_k} \tag{16}$$

ergeben. Es versteht sich wohl von selbst, dass die Matrix C^{-1} nicht explizit gebildet werden darf, da sonst die schwache Besetzung der gegebenen Matrizen A und B vollständig verloren gehen würde.

Der Lanczos-Algorithmus für $C^{-1} = L^T(A - \mu B)^{-1}L$ lautet nach einer Ergänzung von (12) formal wie folgt:

<u>Start</u>: Wähle μ und $\underline{r}_o \neq \underline{0}$; setze $\underline{q}_o = \underline{0}$, $\beta_o = \|\underline{r}_o\|$.	
<u>Iteration</u>: (für j = 1,2,3,...)	
1. $\underline{q}_j = \underline{r}_{j-1} / \beta_{j-1}$ (Normierung)	
2. $\underline{u}_j = L^T(A - \mu B)^{-1}L\ \underline{q}_j$	
3. $\underline{r}_j = \underline{u}_j - \beta_{j-1}\,\underline{q}_{j-1}$ (Orth. bez. $\underline{q}_{j-1}$)	
4. $\alpha_j = \underline{q}_j^T\,\underline{r}_j$	(17)
5. $\underline{r}_j := \underline{r}_j - \alpha_j\,\underline{q}_j$ (Orth. bez. $\underline{q}_j$)	
6. ev. Nachorthogonalisierung von $\underline{r}_j$.	
7. $\beta_j = \|\underline{r}_j\|$	
8. ev. Ritzwerte/-vektoren aus T_j, Q_j.	
Test (Konvergenz Ritzwerte/-vektoren)	

Die Berechnung von $\underline{u}_j$ aus $\underline{q}_j$ im zweiten Schritt von (17) erfordert einerseits die Cholesky-Zerlegung der positiv definiten Matrix $B = L\,L^T$, wobei die Hüllenstruktur von B ausgenützt werden kann, und anderseits die Lösung eines linearen Gleichungssystems mit der zwar symmetrischen, aber im allgemeinen indefiniten Matrix $A - \mu B$.

Der Lanczos-Algorithmus (17) mit zwei erforderlichen, speicheraufwendigen Matrixzerlegungen ist für die Rechenpraxis noch nicht allzu geeignet. Mit Hilfe einiger einfacher Modifikationen kann die Cholesky-Zerlegung von $B = L\,L^T$ tatsächlich wieder eliminiert werden. Dazu multiplizieren wir die 1., 2., 3. und 5. Gleichung in (17) von links mit L^{-T} und definieren die neuen Grössen

$$L^{-T}\underline{q}_j =: \tilde{\underline{q}}_j,\quad L^{-T}\underline{r}_j =: \tilde{\underline{r}}_j,\quad L^{-T}\underline{u}_j =: \tilde{\underline{u}}_j. \tag{18}$$

Für die skalaren Grössen β_j und α_j resultieren mit (18) die Darstellungen

$$\beta_j = (\underline{r}_j^T\,\underline{r}_j)^{\frac{1}{2}} = (\tilde{\underline{r}}_j^T\, L\, L^T\, \tilde{\underline{r}}_j)^{\frac{1}{2}} = (\tilde{\underline{r}}_j^T\, B\, \tilde{\underline{r}}_j)^{\frac{1}{2}} =: \|\tilde{\underline{r}}_j\|_B, \tag{19}$$

$$\alpha_j = \underline{q}_j^T\,\underline{r}_j = \tilde{\underline{q}}_j^T\; L\, L^T\, \tilde{\underline{r}}_j = \tilde{\underline{q}}_j^T\, B\, \tilde{\underline{r}}_j = \tilde{\underline{r}}_j^T\, B\, \tilde{\underline{q}}_j. \tag{20}$$

Nach entsprechender Substitution in der 2. Gleichung von (17) erhält der Lanczos-Algorithmus für das inverse, spektralverschobene Eigenwertproblem die Formulierung [11]:

<u>Start</u>: Wähle μ und $\tilde{\underline{r}}_o$; setze $\tilde{\underline{q}}_o = \underline{0}$.

$\beta_o = \|\tilde{\underline{r}}_o\|_B = [\tilde{\underline{r}}_o^T\,(B\,\tilde{\underline{r}}_o)]^{\frac{1}{2}}$.

<u>Lanczos-Schritte</u>: (für j = 1,2,3,...)

1. $\tilde{\underline{q}}_j = \tilde{\underline{r}}_{j-1} / \beta_{j-1}$ (Normierung)
2. $(A - \mu B)\,\tilde{\underline{u}}_j = B\,\tilde{\underline{q}}_j$
3. $\tilde{\underline{r}}_j = \tilde{\underline{u}}_j - \beta_{j-1}\,\underline{q}_{j-1}$ (B-Orth. bez. $\tilde{\underline{q}}_{j-1}$)
4. $\alpha_j = \tilde{\underline{r}}_j^T\,(B\,\tilde{\underline{q}}_j)$ (21)
5. $\tilde{\underline{r}}_j := \tilde{\underline{r}}_j - \alpha_j\,\tilde{\underline{q}}_j$ (B-Orth. bez. $\tilde{\underline{q}}_j$)
6. ev. Nachorthogonalisierung von $\tilde{\underline{r}}_j$
7. $\beta_j = \|\tilde{\underline{r}}_j\|_B = [\underline{r}_j^T\,(B\,\tilde{\underline{r}}_j)]^{\frac{1}{2}}$
8. Ritzwerte aus T_j : $\rho_1, \rho_2, \ldots, \rho_j$

   ev. Ritzvektoren aus $\tilde{Q}_j$

   Test auf Abbruch

Der Rechenaufwand des Algorithmus (21) setzt sich wie folgt zusammen. In jedem Lanczos-Schritt ist ein lineares Gleichungssystem mit der konstanten, von μ abhängigen Systemmatrix $F := A - \mu B$ zu lösen. Nach einmaliger, geeigneter Zerlegung der Matrix F sind die Prozesse des Vorwärts- und Rückwärtseinsetzens erforderlich. Eine erste Multiplikation eines Vektors mit der

Matrix B ist nur nötig zur Berechnung des Hilfsvektors $\underline{h} = B\,\tilde{\underline{r}}_j$, wie er zur Bildung von β_j gebraucht wird. Da aber im Schritt 1 der Vektor $\tilde{\underline{q}}_j$ durch Normierung aus $\tilde{\underline{r}}_{j-1}$ hervorgeht, kann der Vektor $B\,\tilde{\underline{q}}_j$ aus $\underline{h}$ durch entsprechende skalare Division gewonnen werden. Eine zweite Matrix-Vektor-Multiplikation mit B ist nur dann erforderlich, falls der Vektor $\tilde{\underline{r}}_j$ im Schritt 6 bezüglich aller vorhergehenden $\tilde{\underline{q}}_i$ nachorthogonalisiert werden soll.

5. Zur praktischen Durchführung

Zur Auflösung des linearen Gleichungssystems mit der symmetrischen, aber im allgemeinen indefiniten Matrix $F = A - \mu B$ benötigen wir eine Zerlegung der Form $F = \tilde{L}\,D\,\tilde{L}^T$. Diagonalstrategie würde zwar die Symmetrie und die Hüllenstruktur bewahren, ist aber aus numerischen Gründen nicht sehr geeignet. Eine zweckmässige Pivotierung zur Bewahrung der Symmetrie besteht in der Anwendung von geeigneten Zeilen- und Kolonnenpermutationen verbunden mit Hilfskongruenztransformationen zur Erzeugung eines Pivotelementes in der Diagonale. Um schliesslich die Hüllenstruktur optimal auszunützen, wird die Pivotierung in jedem Schritt auf eine (3×3)-Untermatrix beschränkt. Bei dieser (3×3)-Pivotierung entstehen zwar Elemente ℓ_{ij} mit $(i,j) \notin \mathrm{Env}(F)$, doch zeigt eine Analyse, dass der Zerlegungsprozess in einer leicht erweiterten Hülle durchführbar ist, in welcher in jeder Zeile höchstens zwei Indexpaare nach links im Vergleich zu Env(F) hinzuzufügen sind [10].

Die problemgerechte Wahl der Spektralverschiebung μ zur sukzessiven Berechnung von bestimmten Gruppen aufeinanderfolgender Eigenwerte ist entscheidend für den Algorithmus (21). Bei unzweckmässiger Festsetzung der Werte μ für die einzelnen Läufe besteht durchaus die Möglichkeit, dass Eigenwerte ausgelassen werden. Dies kann etwa dann eintreten, falls der nächste Wert μ nach einem festen Gesetz auf Grund der bereits berechneten Eigenwerte bestimmt wird, wie dies in [1,2] vorgeschlagen ist. Eine durchaus praktikable Lösung der genannten Schwierigkeit besteht darin, mit Hilfe der Bisektionsmethode Intervalle derart zu bestimmen, dass sie je ungefähr die gewünschte Zahl von untereinander verschiedenen, gesuchten Eigenwerten enthalten. Die Mittelpunkte der so ermittelten Intervalle stellen geeignete μ-Werte dar. Der oben erwähnte Zerlegungsalgorithmus bildet die Grundlage dazu.

Sollen pro Lanczos-Lauf nur wenige Eigenwerte in der Nähe von μ bestimmt werden, so ist die dazu erforderliche Zahl von Lanczos-Schritten klein. Auf die Nachorthogonalisierung der Basisvektoren kann somit verzichtet werden, falls gleichzeitig die Zahl der Lanczos-Schritte beschränkt wird.

Die Berechnung der Eigenvektoren erfolgt üblicherweise auf Grund von (11) durch Bildung der Ritzvektoren, d.h. durch Rücktransformation der Basisvektoren. Das liefert aber nur dann brauchbare Näherungen, falls der Prozess der Nachorthogonalisierung angewandt worden ist. Wegen (14), (15) und (18) ergeben sich aber aus den Basisvektoren $\tilde{\underline{q}}_j$, wie sie aus dem Algorithmus (21) resultieren, direkt Näherungen der Eigenvektoren $\underline{x}_k$ von (1). Es ist somit auch an dieser Stelle keine weitere Rücktransformation mit der Matrix L der Cholesky-Zerlegung von B mehr nötig.

Ein anderer Zugang, die Eigenvektoren von (1) zu berechnen, besteht in der inversen Vektoriteration, wie sie in der Bisektionsmethode verwendet wird. Gute Näherungen $\bar{\lambda}$ von Eigenwerten sind ja durch die Ritzwerte bekannt, so dass die Vektorfolge $\underline{z}^{(\nu)}$ aus der Vorschrift

$$(A - \bar{\lambda}B)\ \underline{z}^{(\nu)} = B\ \underline{z}^{(\nu-1)}, \quad \nu = 1,2,... \qquad (22)$$

sehr rasch gegen die Richtung des Eigenvektors $\underline{x}_k$ zum Eigenwert λ_k konvergiert, der am nächsten zu $\bar{\lambda}$ ist. Die in (22) benötigte weitere Zerlegung von $(A - \bar{\lambda}B) = \tilde{L}\ D\ \tilde{L}^T$ liefert zudem die Information über den Index k des zugehörigen Eigenwertes und damit eine Kontrollmöglichkeit über die Vollständigkeit der berechneten Eigenwerte. Weiter lässt sich die Näherung $\bar{\lambda}$ mit Hilfe des Rayleighschen Quotienten auf diese Weise verbessern [8]. Schliesslich erlaubt das Vorgehen, einen fälschlicherweise als Näherung eines Eigenwertes angesehenen Wert $\bar{\lambda}$ zu erkennen auf Grund zu langsamer Konvergenz der Iterierten $\underline{z}^{(\nu)}$.

Das Abbruchkriterium besteht darin festzustellen, ob die gewünschte Anzahl unter den absolut grössten Ritzwerten ρ_i unter einer vorgegebenen Toleranz konvergiert sind. Wir haben diesen Abbruchtest verwendet, weil das übliche Kriterum auf Grund des Residuums in bestimmten technischen Anwendungen versagt.

Der Startvektor $\tilde{\underline{r}}_o$ für einen Lanczos-Lauf wird zufällig festgesetzt. Sind bereits einige Eigenvektoren $\underline{x}_i$ von (1) berechnet worden, empfiehlt es sich, $\tilde{\underline{r}}_o$ B-orthogonal zu den bekannten $\underline{x}_i$ zu wählen, damit die schon bestimmten Eigenwerte theoretisch ausgeschlossen werden. Da aber nur Näherungen für die $\underline{x}_i$ vorliegen, können durchaus in einem späteren Lauf schon bekannte Eigenwerte nochmals berechnet werden. Diese sind selbstverständlich zu eliminieren. Besteht aber die Möglichkeit von mehrfachen Eigenwerten, kann dieser Fall mit Zusatzmassnahmen relativ einfach erfasst werden. Wegen der Irreduzibilität von T_j kann ein einzelner Lanczos-Lauf im Prinzip keine mehrfachen Eigenwerte liefern. Es sind entsprechend viele Läufe dazu erforderlich.

Für eine erfolgreiche Anwendung des Verfahrens sind noch eine Reihe von weiteren Detailfragen zu berücksichtigen. Man vergleiche dazu [11].

6. Numerische Beispiele

Die drei Beispiele von Abschnitt 1 sind zu Vergleichszwecken mit verschiedenen Verfahren behandelt worden. In allen Fällen waren die p = 12 kleinsten Eigenwerte mit den zugehörigen Eigenvektoren gesucht. Die Eigenwerte sollten mit einer Genauigkeit von 10 wesentlichen Stellen und die Eigenvektoren mit einer absoluten Genauigkeit von 5 Stellen nach dem Komma unter der Normierung $\| \underline{x}_k \|_\infty = 1$ bestimmt werden. Als erste Methode wurde die simultane inverse Vektoriteration (SIVIT) angewandt, wobei mit $p^* = 14$ gleichzeitig iterierten Vektoren gearbeitet wurde. Weiter wurde die Bisektionsmethode (BISECT) verwendet. In beiden Fällen wurden die Rechenprogramme aus [9] benutzt, und die Rechnungen erfolgten auf der IBM 3033 der Universität Zürich. Schliesslich wurden die gewünschten Ergebnisse mit verschiedenen Varianten des Lanczos-Algorithmus berechnet. Die Zahl der Lanczos-Schritte pro Lauf war in allen Fällen, wo nichts anderes vermerkt ist, auf 20 beschränkt. Ritzwerte wurden als Näherungen für Eigenwerte akzeptiert, sobald sich vier wesentliche Ziffern nicht mehr änderten, falls die Eigenvektoren mit inverser Vektoriteration bestimmt werden. Andernfalls mussten zehn wesentliche Ziffern stehen. Die Rechenzeiten sprechen eindeutig zugunsten der Lanczos-Verfahren.

Beispiel 1. (Autoproblem)

Figur 5 Eigenwerte des Autoproblems

a) SIVIT: Bei 96 Iterationsschritten war die Rechenzeit 288 sec.

b) BISECT: Mit dem Startintervall [-0.01,1.0] wurden 64 Zerlegungen und 79 inverse Vektoriterationsschritte bei einer Rechenzeit von 121 sec benötigt.

c) LANCZOS + BISECT: Keine Nachorthogonalisierung, Berechnung der Eigenvektoren durch inverse Vektoriteration, 6 Eigenwerte pro Lanczos-Lauf. Die Rechenzeit betrug 57 sec. Die Tabelle gibt Aufschluss über den Ablauf.

Lauf:	1	2	3
Schritte:	17	9	8
erhaltene Eigenwerte:	$\lambda_1, \ldots, \lambda_7$	$\lambda_8, \lambda_9, \lambda_{10}$	$\lambda_{11}, \lambda_{12}, (\lambda_{13}, \lambda_{14})$

d) LANCZOS: Bestimmung von μ nach [1], keine Nachorthogonalisierung, Berechnung der Eigenvektoren durch inverse Vektoriteration, 4 Eigenwerte pro Lanczos-Lauf. Die Rechenzeit war 52 sec.

Lauf:	1	2	3
Schritte:	18	12	14
erhaltene Eigenwerte:	$\lambda_1,\ldots,\lambda_4$	$\lambda_5,\lambda_7,\ldots,\lambda_{10}$	$\lambda_6,\lambda_{11},\lambda_{12},(\lambda_{14},\lambda_{15})$

Beispiel 2. (Bürohaus)

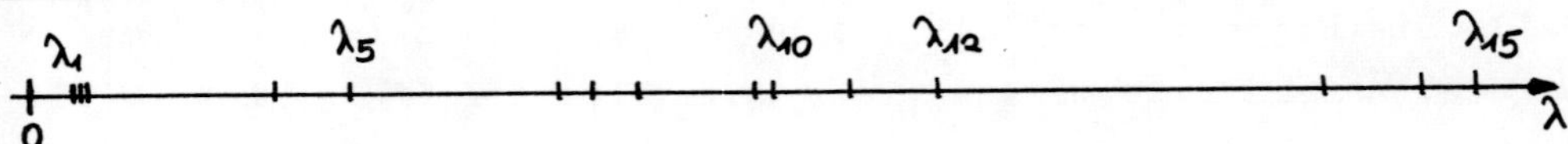

Figur 6 Eigenwerte für Bürohaus

a) SIVIT: Bei 51 Iterationsschritten betrug die Rechenzeit 468 sec.

b) BISECT: Für das Startintervall $[0,10^5]$ waren 59 Zerlegungen und 75 inverse Vektoriterationsschritte bei einer Rechenzeit von 1020 sec erfroderlich.

c) LANCZOS + BISECT: Keine Nachorthogonalisierung, Berechnung der Eigenvektoren durch inverse Vektoriteration, 6 Eigenwerte pro Lanczos-Lauf. Die Rechenzeit war 339 sec.

Lauf:	1	2
Schritte:	13	13
erhaltene Eigenwerte:	$\lambda_1,\ldots,\lambda_5$	$\lambda_6,\ldots,\lambda_{12}$

d) LANCZOS: Bestimmung von μ nach [1], keine Nachorthogonalisierung, Berechnung der Eigenvektoren durch inverse Vektoriteration, 6 Eigenwerte pro Lanczos-Lauf. Die Rechenzeit betrug 325 sec.

Lauf:	1	2	3
Schritte:	15	15	8
erhaltene Eigenwerte:	$\lambda_1,\ldots,\lambda_6$	$\lambda_9,\ldots,\lambda_{14}$	λ_7,λ_8

e) LANCZOS: Mit Nachorthogonalisierung, Berechnung der Eigenvektoren durch Rücktransformation, 12 Eigenwerte pro Lanczos-Lauf.

Mit 29 Schritten wurden in einem einzigen Lauf alle Eigenwerte bestimmt in der sehr kurzen Rechenzeit von 78 sec. Sie erklärt sich dadurch, dass infolge der grossen Bandbreite der Matrizen A und B die Zerlegung von F = A - μB sehr rechenintensiv ist.

Beispiel 3. (Hochspannungsmast)

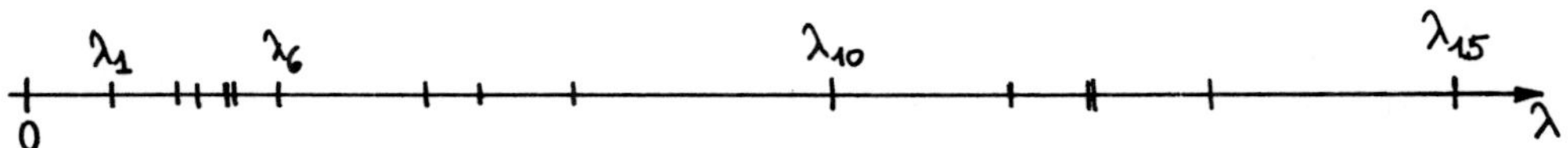

Figur 7 Eigenwerte für Hochspannungsmast

a) SIVIT: Bei 81 Iterationsschritten war die Rechenzeit 340 sec.

b) BISECT: Mit dem Startintervall $[0,10^7]$ lieferten 62 Zerlegungen und 87 inverse Vektoriterationsschritte die Resultate mit einer Rechenzeit von 252 sec.

c) LANCZOS + BISECT: Keine Nachorthogonalisierung, Berechnung der Eigenvektoren durch inverse Vektoriteration, 6 Eigenwerte pro Lanczos-Lauf. Die Rechenzeit war 118 sec.

Lauf:	1	2
Schritte:	13	18
erhaltene Eigenwerte:	$\lambda_1,\ldots,\lambda_6$	$\lambda_7,\ldots,\lambda_{12},(\lambda_{13},\lambda_{14})$

d) LANCZOS: Bestimmung von μ nach [1], keine Nachorthogonalisierung, Berechnung der Eigenvektoren durch inverse Vektoriteration, 6 Eigenwerte pro Lanczos-Lauf. Die Rechenzeit war 125 sec.

Lauf:	1	2	3	4	5
Schritte:	17	14	6	6	6
erhaltene Eigenwerte:	$\lambda_1,\ldots,\lambda_7$	$\lambda_8,\ldots,\lambda_{11}$	λ_{18}	λ_{14}	λ_{12}

e) LANCZOS: Mit Nachorthogonalisierung, Berechnung der Eigenvektoren durch Rücktransformation, 12 Eigenwerte pro Lanczos-Lauf. In einem einzigen Lauf von 36 Schritten wurden alle Eigenwerte in der Rechenzeit von 72 sec berechnet.

Abschliessend muss zu den angeführten Rechenzeiten auf die besondere Charakteristik der verwendeten Rechenanlage hingewiesen werden. Die IBM 3033 arbeitet auf der Basis eines virtuellen Speichers von fast unbegrenzter Kapazität, und der Datentransfer (paging) erfolgt automatisch.

Verdankung. An dieser Stelle danke ich Herrn P. Waldvogel für die Durchführung der umfangreichen numerischen Experimente.

LITERATUR

[1] Ericsson, T.: Implementation and Applications of the Spectral Transformation Lanczos Algorithm. In Kogström, B., Ruhe, A.: Matrix Pencils. Springer, New York 1983.

[2] Ericsson, T., Ruhe, A.: The Spectral Transformation Lanczos Method for the Numerical Solution of Large Sparse Generalized Eigenvalue Problems. Math. Comp. 35(1980), 1251-1268.

[3] Lanczos, C.: An Iteration Method for the Solution of the Eigenvalue Problem of Linear Differential and Integral Operators. J. Res. Nat. Bur. Standards. Sect. B. 45(1950), 225-280.

[4] Paige, C.C.: The Computation of Eigenvalues and Eigenvectors of Very Large Sparse Matrices. Ph. D. thesis, University of London, 1971.

[5] Paige, C.C.: Computational Variants of the Lanczos Method for the Eigenproblem. J. Inst. Math. Appl. 10(1972), 373-381.

[6] Paige, C.C.: Error Analysis of the Lanczos Algorithm for Tridiagonalizing a Symmetric Matrix. J. Inst. Math. Appl. 18(1976), 341-349.

[7] Parlett, B.N.: The Symmetric Eigenvalue Problem. Englewood Cliffs, N.J., Prentice Hall 1980.

[8] Schwarz, H.R.: Methode der finiten Elemente. Stuttgart, Teubner 1980.

[9] Schwarz, H.R.: FORTRAN-Programme zur Methode der finiten Elemente. Stuttgart, Teubner 1981.

[10] Waldvogel, P.: Bisection for $Ax = \lambda Bx$ with Matrices of Variable Band Width. Computing 28(1982), 171-180.

[11] Waldvogel, P.: Numerische Behandlung von allgemeinen Eigenwertproblemen. Dissertation, Zürich (wird 1984 erscheinen).

Prof. Dr. H.R. Schwarz
Universität Zürich
Rämistrasse 74
CH - 8001 Zürich/Schweiz

ISNM, Vol. 69
Numerical Treatment of
Eigenvalue Problems, Vol. 3

EFFICIENT BRANCH SWITCHING IN SYSTEMS OF NONLINEAR EQUATIONS

R. Seydel

A numerical method is proposed for branch switching in nonlinear equations. The method consists of two steps. The first step calculates a rough approximation of a solution on an emanating branch. The second step yields the accurate approximation by solving a special equation with certain selectivity properties. The procedure is very efficient because neither the exact location of the bifurcation point nor directions of branches need to be calculated.

1. Introduction

Consider a system of nonlinear equations

(1) $g(x,\lambda) = 0$,

where $x \in \mathbb{R}^n$, $\lambda \in \mathbb{R}$, $g\colon D \subset \mathbb{R}^{n+1} \longrightarrow \mathbb{R}^n$ sufficiently smooth. Suppose that there are two branches of solutions $\{(x,\lambda)\}$, $\{(z,\lambda)\}$ of (1) intersecting in a simple bifurcation point (x_0,λ_0) , $z \in \mathbb{R}^n$, $x_0 \in \mathbb{R}^n$. Assume further that the branch $x(\lambda)$ can be parametrized by λ. The problem considered in the present contribution is the branch switching. Methods for switching branches in nonlinear equations were published, for example, by Keller [1] and Kubíček, Klíč [2].

In the following, an algorithm will be developed for approximating a solution on the emerging branch $\{(z,\lambda)\}$, starting from $x(\lambda)$. The proposed method will be similar to a method for handling the case of ordinary differential equations [6]. A characteristic feature of the present device is its efficiency: it is not necessary to approximate the bifurcation point. We only assume that (at least) one solution $\bar{x}=x(\bar{\lambda})$ is calculated "not far" from the bifurcation point.

2. The principle

Before going into details, we outline the essential steps (illustrated by Fig.1) of a method for switching branches. The k-th component of x and z is denoted by x_k, z_k $(1\leqslant k\leqslant n)$.

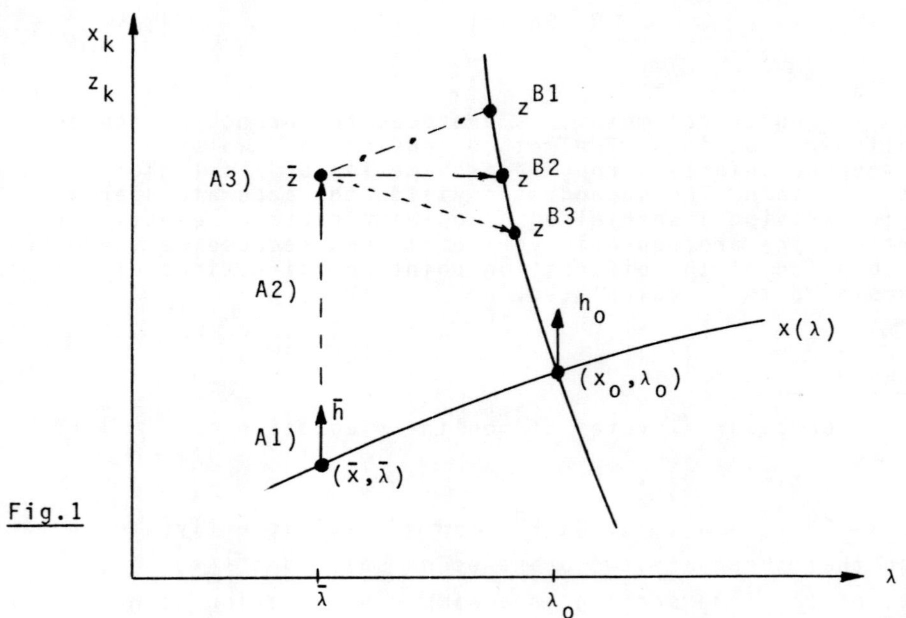

Fig.1

step A : rough approximation $\bar{z}\approx z$

A1) calculation of direction vector $\bar{h}\approx\frac{z-x}{\|z-x\|}$

A2) fix a distance $\|x-z\|=\delta$

A3) $\bar{z} := \bar{x} + \delta\bar{h}$

step B : accurate approximation of z , starting from $\bar{z}$

Solve an appropriate equation $G(X)=0$ with better selective properties than $g(x,\lambda)=0$: $\bar{z}\rightarrow z$, $\bar{z}\not\rightarrow x$ (one of at least three possibilities B1, B2, B3).

In step A, the calculations are based on the data that were obtained during the calculation of $\bar{x}$. After having carried out step A, one might succeed in approximating a solution z by solving equation (1), using $\bar{z}$ for initial guess. However, in difficult cases, convergence might yield the "known" branch $x(\lambda)$ instead of z . Thus, in step B, a slightly different equation is solved which has better selective properties *).

3. The algorithm, based on one solution

During the approximation of a solution of (1) by, for example, Newton's method, the matrix of the derivatives

$$Z(x,\lambda) := g_x(x,\lambda)$$

is used. Thus, apart from $\bar{x}=x(\bar{\lambda})$, an approximation of $Z(\bar{x},\bar{\lambda})$ can be considered to be available. As will be shown in the sequel, everything we need is contained in this matrix.

Definitions:

e_ℓ : ℓ-th unit vector of $\mathbb{R}^n$

$$Z_{\ell k}(x,\lambda) := (I-e_\ell e_\ell^T)Z(x,\lambda) + e_\ell e_k^T$$

(2) $$h(x,\lambda) := Z_{\ell k}(x,\lambda)^{-1}e_\ell$$

$$h_o := h(x_o,\lambda_o)$$

$$\bar{h} := h(\bar{x},\bar{\lambda})$$

$Z_{\ell k}$ is the matrix Z where the ℓ-th row $(1\leq\ell\leq n)$ is replaced by the k-th unit vector. The n-vectors h_o , $\bar{h}$ are solutions of linear equations. $\bar{h}$ is a vector in the direction of the unknown solution z ,

$$\delta\bar{h} \approx z-\bar{x} \quad ,$$

---

*) The term "selective" reminds of the paper of Scheurle [4] where a selective iteration is established which converges to a solution stable in a physical sense. The "selectivity" in the present paper is simply a technical condition without physical meaning.

for some $\delta \in \mathbb{R}$. In fact, under certain assumptions (essentially g is of class C^1 and $Z(x_0,\lambda_0)$ has a one dimensional null-space) two properties hold that can be stated in simplified form as

$$(\bar{x},\bar{\lambda}) \approx (x_0,\lambda_0) \quad \text{implies}$$

(3a) $\bar{h} \approx h_0$ (from [5])

(3b) $z-\bar{x} \approx \delta h_0$ for some δ (from [6]).

The z-branch may be defined on one "side" of λ_0 only; the case that z and $\bar{x}$ are situated on different sides is excluded for the moment.

By (3) one realizes that

$$\bar{z} := \bar{x} + \delta\bar{h} \approx \bar{x} + \delta h_0 \approx z \quad . \tag{4}$$

That is to say, for some appropriate value of δ, $(\bar{z},\bar{\lambda})$ is a rough approximation of a solution $(z,\bar{\lambda})$ of (1). However, for given $\bar{\lambda}$, a corresponding value of δ is unknown. We therefore fix the "distance" δ and calculate λ in dependence of δ, $\lambda=\lambda(\delta)$. With this approach, z and $\bar{x}$ are allowed to be situated on different "sides" of λ_0. - There are at least three possibilities for the calculation of $\lambda(\delta)$, $z(\delta)$:

<u>First possibility (B1)</u>: x and z are calculated in one equation simultaneously. Denoting $X:=(x,\lambda,z)^T$, the equation

$$G^{B1}(X,\delta) := \begin{pmatrix} g(x,\lambda) \\ g(z,\lambda) \\ z_k-x_k-\delta \end{pmatrix} = 0 \tag{5}$$

is solved, starting from the initial guess $(\bar{x},\bar{\lambda},\bar{z})^T$. Solving (5) is selective because for $\delta \neq 0$ one obtains via $X(\delta)$ two solutions (x,λ), (z,λ) being situated on different branches. This first possibility of parallel computation is rather expensive because of its double dimension of $2n+1$.
The following simplified version is preferable:

Second possibility (B2): Solve the equation

$$(6) \qquad G^{B2}(X,\delta) := \begin{pmatrix} g(z,\lambda) \\ z_k - \bar{z}_k \end{pmatrix} = 0$$

with $X:=(z,\lambda)^T \in \mathbb{R}^{n+1}$, starting from the initial guess $(\bar{z},\bar{\lambda})^T$. The dependence on the parameter δ is contained in the last component, $\bar{z}_k = \bar{z}_k(\delta)$, see (4). For the geometrical meaning compare Fig.1.

Third possibility (B3): The third possibility exploits different symmetry properties on different branches. Without presenting details, we just give the principle. With $X:=(z,\lambda)^T$ one solves, for example, the equation

$$(7) \qquad G^{B3}(X,\delta) := \begin{pmatrix} g(z,\lambda) \\ (z_i - z_j) - (\bar{z}_i - \bar{z}_j) \end{pmatrix} = 0 \quad .$$

Equation (7) means the case that $x_i = x_j$ holds for the $x(\lambda)$-branch. If this property is lost by the z-branch (recognizable by $\bar{z}$ from (4)) then the z-branch can be parametrized by $z_i - z_j = \rho$, $\rho \in \mathbb{R}$. Solving (7) is selective in such a way that symmetric solutions in the sense $x_i = x_j$ are excluded.

Summary: Let $(\bar{x},\bar{\lambda})$ be a solution of (1) not far from a bifurcation point. Fix a $\delta \neq 0$ by

$$(8) \qquad \delta := \delta_0 \cdot \max\{1, |\bar{x}_k|\} \quad , \; \delta_0 = 0.02 \text{ , for example.}$$

Then calculate the vector $\bar{h}$ by solving the linear equation in (2) which is based on the values of $g_x(\bar{x},\bar{\lambda})$ and the choice of indices ℓ and k. By (4) one obtains a rough approximation $\bar{z}$ for the emanating branch. z is determined by solving equation (6) or (7); in a few cases equation (5) may be used.

4. The algorithm, based on two solutions

In the foregoing section, formula have been given which are based on one solution $(\bar{x},\bar{\lambda})$ only. In practice, one calculates solutions in homotopy chains. Especially in detecting bifurcations, normally two solutions $(\bar{x},\bar{\lambda})$, $(\bar{\bar{x}},\bar{\bar{\lambda}})$ are needed in order to be able to decide whether there is a bifurcation nearby. This decision is based on evaluating a certain test-function $\tau(x,\lambda)$ with property $\tau(x_0,\lambda_0)=0$. A testfunction suitable for our method was introduced in [5],

$$\tau(x,\lambda) := e_\ell^T Z(x,\lambda) Z_{\ell k}(x,\lambda)^{-1} e_\ell = e_\ell^T Z(x,\lambda) h(x,\lambda) \quad . \tag{9}$$

If the vector h is calculated, the evaluation of τ means just evaluating the scalar product of h and the ℓ-th row of $g_x(x,\lambda)$.

Based on two solutions $(\bar{x},\bar{\lambda})$, $(\bar{\bar{x}},\bar{\bar{\lambda}})$, the values of $\bar{h}$, $\bar{\bar{h}}$, $\tau(\bar{x},\bar{\lambda})$, $\tau(\bar{\bar{x}},\bar{\bar{\lambda}})$, $\bar{z}$ and $\bar{\bar{z}}=\bar{\bar{x}}+\delta\bar{\bar{h}}$ can be easily calculated by means of the algorithm of the preceding section. In view of this situation, it suggests itself to compute an initial guess for the solution of $G(X,\delta)=0$ that is closer to the bifurcation point. Interpolation or extrapolation to the zero of $\tau(x,\lambda)$ yields the formulas

$$\xi := \tau(\bar{x},\bar{\lambda}) \,/\, (\tau(\bar{x},\bar{\lambda})-\tau(\bar{\bar{x}},\bar{\bar{\lambda}}))$$

$$\begin{aligned} \bar{\lambda}_0 &:= \bar{\lambda} + \xi\cdot(\bar{\bar{\lambda}}-\bar{\lambda}) \\ \bar{x}_0 &:= \bar{x} + \xi\cdot(\bar{\bar{x}}-\bar{x}) \\ \bar{h}_0 &:= \bar{h} + \xi\cdot(\bar{\bar{h}}-\bar{h}) \\ \bar{z}_0 &:= \bar{z} + \xi\cdot(\bar{\bar{z}}-\bar{z}) \quad . \end{aligned} \tag{10}$$

The initial guess for solving $G(X,\delta)=0$ now consists of the left-hand variables in (10). $(\bar{x}_0,\bar{\lambda}_0)$ serves as approximation of the bifurcation point.

<u>Automation</u>: It is very easy to implement and automate the above procedure of calculating a solution on an emanating branch during homotopy. After each homotopy step the values of h and τ are calculated, yielding pairs

$$(\lambda_1,\tau_1) \;,\; (\lambda_2,\tau_2) \;,\; (\lambda_3,\tau_3) \;,\; \ldots\ldots \quad .$$

A change of sign of τ controls an interruption of the continuation for an intermediate solving of $G(X,\delta)=0$. Note that $\tau(x,\lambda)$ may possess poles, the corresponding changes of sign have to be neglected. If the stepsizes $|\lambda_i-\lambda_{i-1}|$ are sufficiently small, a "continuous" change of sign can be detected by criterion (11) which is based on three values τ_i, τ_{i-1}, τ_{i-2} ,

(11) $\quad \tau_i\tau_{i-1} < 0$ <u>and</u> $|\tau_{i-1}| < |\tau_{i-2}|$.

5. An example

We consider a model of a trimolecular reaction in three interconnected reaction cells which is related to the "Brusselator" [3, 2]:

$$\begin{aligned}
2 - 7x_1 + x_1^2x_2 + \lambda(x_3-x_1) &= 0\\
6x_1 - x_1^2x_2 + 10\lambda(x_4-x_2) &= 0\\
2 - 7x_3 + x_3^2x_4 + \lambda(x_1+x_5-2x_3) &= 0\\
6x_3 - x_3^2x_4 + 10\lambda(x_2+x_6-2x_4) &= 0\\
2 - 7x_5 + x_5^2x_6 + \lambda(x_3-x_5) &= 0\\
6x_5 - x_5^2x_6 + 10\lambda(x_4-x_6) &= 0
\end{aligned}$$

This problem of type $g(x,\lambda)=0$, $x\in \mathbb{R}^6$, will be investigated for parameter values $0<\lambda<5$. By means of the continuation algorithm [7] solutions are calculated in chains. Thereby, branch switching is carried out automatically using the method established in the present contribution. In Fig.2, the obtained results are plotted in a $x_1(\lambda)$-branching diagram. As can be seen by Fig.2, there are a lot of branches, branch points are marked. Part of this diagram is enlarged in Fig.3. As expected, numerical computations show that the system $G^{B1}(X,\delta)$ of the double dimension (5) is about four times as slow as one of the smaller systems (6) or (7).

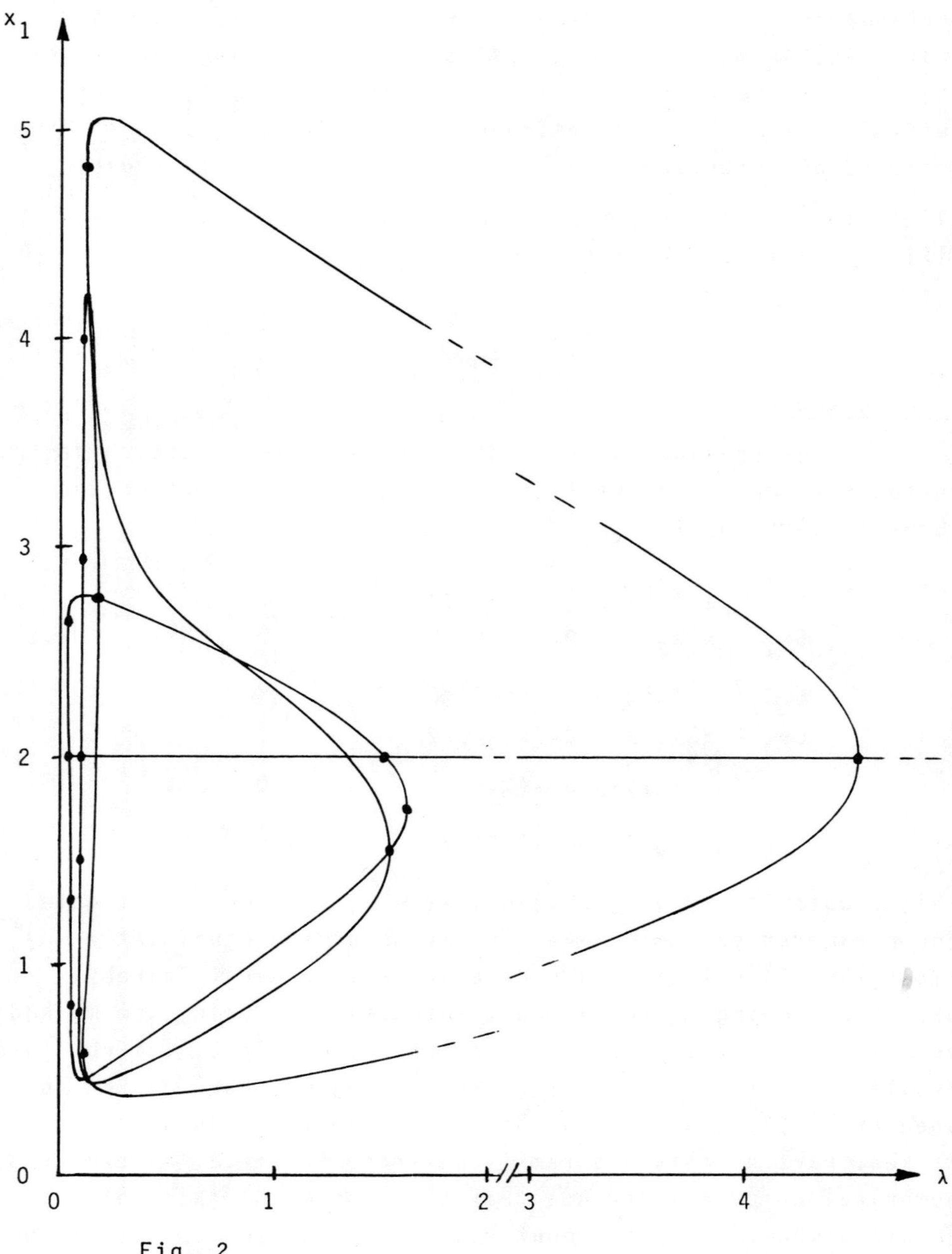

Fig. 2

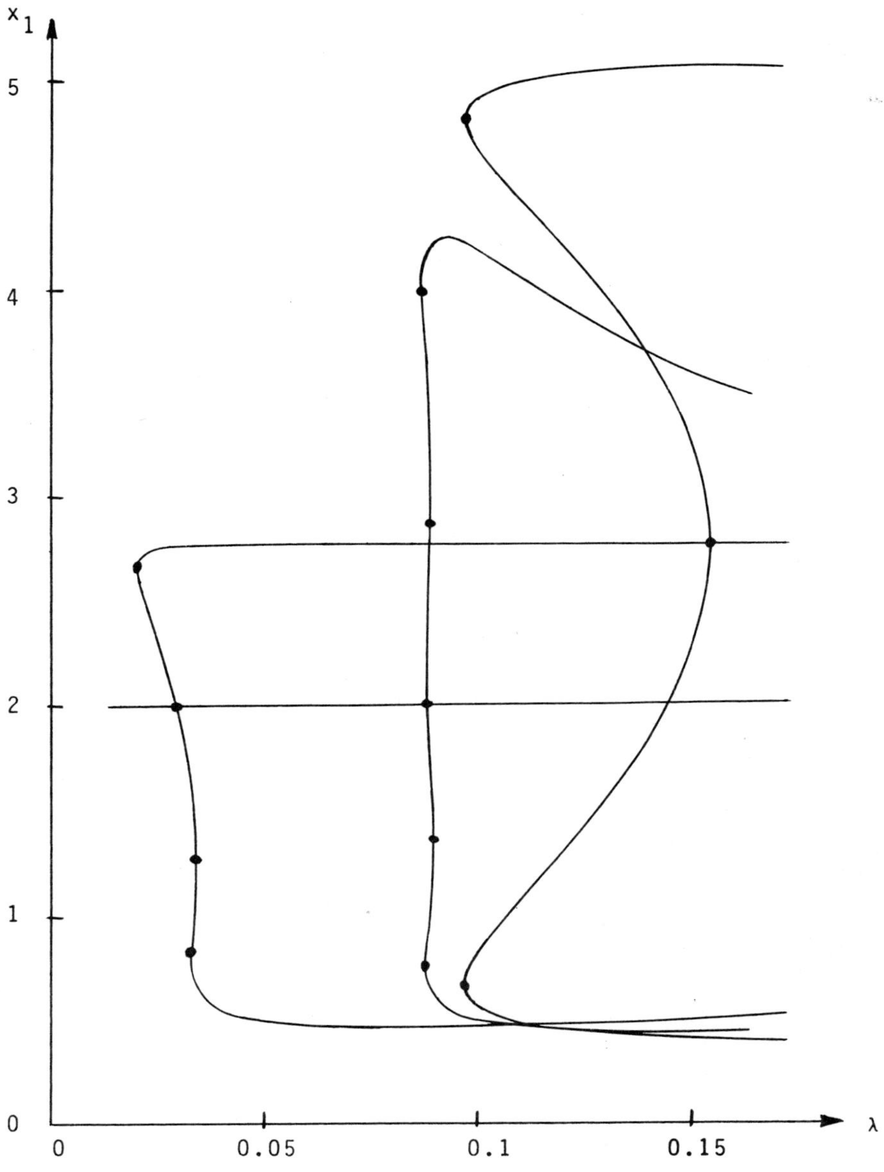

Fig. 3

<u>Appendix</u>: Although not needed for the present method of branch switching, we complete our numerical results by listing accurate values of all branch points (6 bifurcation and 10 turning points) in the final table. These branch points have been calculated by the method of [5].

λ_o	x_1
0.0203067	2.64153
0.0295552	2.00000
0.0335279	0.838479
0.0341459	1.265004
0.0872084	0.764918
0.0872084	3.98597
0.0886656	2.00000
0.0893470	1.35524
0.0893470	2.87172
0.0976019	4.80000
0.0976019	0.653982
0.155218	2.75475
1.50378	2.00000
1.53561	1.57162
1.60171	1.76993
4.51133	2.00000

<u>Table</u> of branch points, rounded to 6 digits

References

1 Keller, H.B.: Numerical solution of bifurcation and nonlinear eigenvalue problems. in: P.H.Rabinowitz: Applications of bifurcation theory. New York, Academic Press 1977.

2 Kubíček, M., Klíč, A.: Direction of branches bifurcating at a bifurcation point. Determination of starting points for a continuation algorithm. Appl. Math. Comput., in press.

3 Glansdorff, P., Prigogine, I.: Thermodynamic theory of structure, stability and fluctuations. London, Wiley-Interscience 1971.

4 Scheurle, J.: Selective iteration and applications. J. Math. Anal. Appl. 59 (1977), 596-616.

5 Seydel, R.: Numerical computation of branch points in nonlinear equations. Numer. Math. 33 (1979), 339-352.

6 Seydel, R.: Neue Methoden zur numerischen Berechnung abzweigender Lösungen bei Randwertproblemen gewöhnlicher Differentialgleichungen. Habilitationsschrift TU München 1981. partly published in: Numer. Math. 41 (1983), 93-116.

7 Seydel, R.: A continuation algorithm with step control. submitted for publication.

Rüdiger Seydel
Mathematisches Institut
TU München
Arcisstr. 21
D-8000 München 2

ISNM, Vol. 69
Numerical Treatment of
Eigenvalue Problems, Vol. 3

UNTERE SCHRANKEN FÜR EIGENWERTE IN DER LINEAREN ELASTIZITÄTSTHEORIE

Waldemar Velte

For an eigenvalue problem occuring in linear elasticity theory, lower bounds to the eigenvalues are constructed by a finite difference method. This method is a variant of those presented earlier by H. F. Weinberger and J. Hersch respectively.

1. Problemstellung

Ein elastischer Körper nehme im Ausgangszustand das Gebiet G ein. Die Verschiebungen der einzelnen Massenpunkte seien durch den Vektor $\underline{u} = (u,v,w)$ beschrieben. Im Falle eines isotropen elastischen Materials führt dann das Problem der Eigenschwingungen des elastischen Körpers bei festgehaltenem Rand auf eine Eigenwertaufgabe (EWA), die sich nach Division durch einen Faktor (Materialkonstante) wie folgt schreiben läßt:

$$\begin{aligned} \Delta \underline{u} + c \operatorname{grad} \operatorname{div} \underline{u} + \lambda \underline{u} &= 0 \quad \text{in } G \\ \underline{u} &= 0 \text{ auf } \partial G \quad . \end{aligned} \tag{1}$$

Dabei ist Δ der Laplaceoperator und c eine positive Konstante, die in einfacher Weise von den Materialkonstanten (Lamésche Konstanten) abhängt. λ ist der Eigenwertparameter ([1] S. 20, [2] S. 90).

Im folgenden interessiert uns vor allem der zweidimensionale Fall, wobei G ein beschränktes, einfach zusammenhängendes Gebiet in der x,y-Ebene mit stückweise glattem Rand ∂G sei. Mit der Notation $\underline{u} = (u,v)$ lautet dann die EWA (1) ausführlich geschrieben:

$$\begin{aligned} u_{xx} + u_{yy} + c\,(u_{xx} + v_{xy}) + \lambda u &= 0 \\ v_{xx} + v_{yy} + c\,(u_{xy} + v_{yy}) + \lambda v &= 0 \qquad (2) \\ u = v = 0 \text{ auf } \partial G \ . \end{aligned}$$

Der zugehörige Rayleighsche Quotient lautet

$$R(\underline{u}) = \int_G (|\mathrm{grad}\,\underline{u}|^2 + c(\mathrm{div}\,\underline{u})^2)\,dx\,dy \Big/ \int_G |\underline{u}|^2 dx\,dy \qquad (3)$$

mit

$$|\mathrm{grad}\,\underline{u}|^2 = u_x^2 + u_y^2 + v_x^2 + v_y^2 \quad , \quad |\underline{u}|^2 = u^2 + v^2 \ .$$

Die EWA besitzt abzählbar unendlich viele Eigenwerte endlicher Vielfachheit,

$$0 < \lambda_1 \leq \lambda_2 \leq \dots \qquad , \quad \lambda_n \to \infty \text{ für } n \to \infty \ ,$$

wobei die Eigenwerte entsprechend ihrer Vielfachheit aufgeführt sind.

Wendet man auf die EWA (2) das Verfahren von Rayleigh und Ritz an, so erhält man bekanntlich obere Schranken für die ersten Eigenwerte ([3],[4],[5]).

Untere Schranken kann man im Prinzip mit der Methode der intermediären Variationsprobleme ([4],[5]) erhalten, indem man die EWA mit $c = 0$ als "Basisproblem" verwendet. In diesem Falle ist das System entkoppelt, und man erhält für u und v getrennt die EWA der eingespannten schwingenden Membran. Allerdings ist die Methode nur für solche Gebiete praktikabel, für die man die Eigenlösungen der schwingenden Membran explizit kennt. Im Sonderfall eines quadratischen Gebietes hat L.Bassotti in [6] untere Schranken mit Hilfe der Fichera'schen Methode der orthogonalen Invarianten berechnet. (Siehe auch [2] S. 90.) Die Schranken sind sehr gut. Allerdings ist der numerische Aufwand erheblich.

Im folgenden soll gezeigt werden, wie man auch bei allgemeineren Gebieten mit Hilfe eines einfachen Differenzenverfahrens zu Näherungen für die ersten Eigenwerte gelangen kann, die in einfacher Weise untere Schranken liefern. Es handelt sich dabei um eine Abwandlung der Methoden, die von H.F. Weinberger [7],[8] und J. Hersch [9],[10] angegeben worden sind.

Bei Weinberger [8] wird hervorgehoben, daß sich seine Methode nicht auf gekoppelte Systeme mit gemischten Ableitungen anwenden läßt. Auch Kuttler [11] betrachtet solche Systeme nicht. Die spezielle Struktur der Gleichungen in (2) erlaubt aber eine Modifikation der ursprünglichen Ideen, die hier zum Ziele führt.

2. Konstruktion unterer Schranken für die Eigenwerte

Die Gleichungen (2) werden in der üblichen Weise mittels finiter Differenzen diskretisiert, wobei allerdings ähnlich wie bei Weinberger [7],[8] ein etwas vergrößertes Gittergebiet zugrundegelegt wird. Wir beschränken uns auf eine Darstellung der Methode im ebenen Fall. Eine Übertragung auf den räumlichen Fall ist möglich.

Es bezeichne Γ_h das quadratische Gitter in der x,y-Ebene mit der Gitterkonstanten $h > 0$ und mit den Gitterpunkten

$$P_{jk} = (x_j, y_k) \quad , \quad x_j = jh \quad , \quad y_k = kh \quad (j,k = 0, \pm 1, \pm 2, \ldots).$$

Auf dem unendlichen Gitter Γ_h werden zwei endliche Teilgitter Γ'_h und Γ''_h betrachtet, die sich wie folgt ergeben: Zunächst werden zu jedem Punkt P_{jk} zwei kleine Parallelogramme der Fläche h^2 konstruiert, nämlich

$$\begin{aligned} Q'_{jk} &= \{(x,y) \mid x = x_j + t \ , \ y = y_k + t + \eta \ , \ 0 < t < h, \ 0 < \eta < h\} \ , \\ Q''_{jk} &= \{(x,y) \mid x = x_j + t + \xi \ , \ y = y_k + t \ , \ 0 < t < h, \ 0 < \xi < h\} \ . \end{aligned} \tag{4}$$

Die Teilgitter werden dann wie folgt definiert:

$$\begin{aligned} \Gamma'_h &= \{P_{jk} \in \Gamma_h \mid Q'_{jk} \cap G \neq \emptyset\} \ , \\ \Gamma''_h &= \{P_{jk} \in \Gamma_h \mid Q''_{jk} \cap G = \emptyset\} \ . \end{aligned} \tag{5}$$

Sie enthalten also zunächst alle Gitterpunkte $P_{jk} \in G$, aber außerdem noch gewisse Punkte außerhalb von G in der Nähe des Randes (Figur).

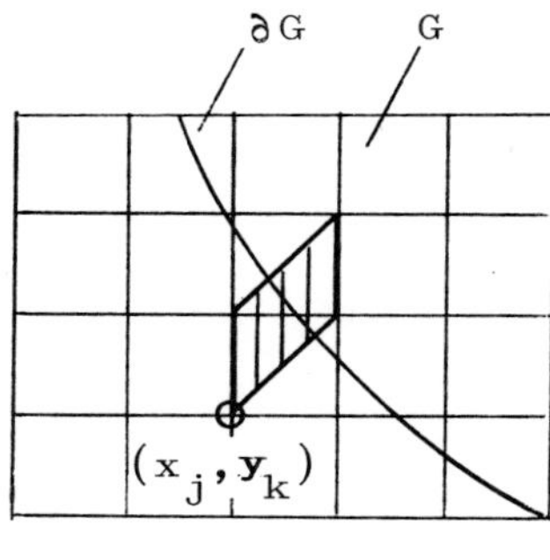

Zu Q'_{jk}

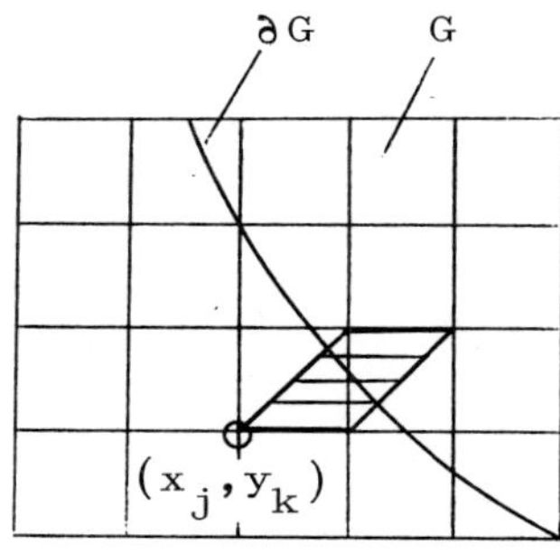

Zu Q''_{jk}

Im folgenden bezeichnet $\underline{U} = (U,V)$ einen Vektor reellwertiger Gitterfunktionen, die formal auf ganz Γ_h betrachtet werden, jedoch mit den Einschränkungen

$$U_{jk} = 0 \quad \text{außerhalb } \Gamma_h' \quad , \qquad V_{jk} = 0 \quad \text{außerhalb } \Gamma_h'' \quad .$$

Das System (2) wird dann wie folgt diskretisiert, wobei ∇_x, ∇_y und $\bar{\nabla}_x, \bar{\nabla}_y$ die vorwärts bzw. rückwärts genommenen Differenzen bezüglich x und y bezeichnen: Der Laplaceoperator wird ersetzt durch

$$\Delta_h U_{jk} = h^{-2}(\bar{\nabla}_x \nabla_x + \bar{\nabla}_y \nabla_y) U_{jk} \sim h^{-2} \begin{bmatrix} & 1 & \\ 1 & -4 & 1 \\ & 1 & \end{bmatrix} U_{jk} \quad .$$

Die Operatoren div und grad werden mit vorwärts- bzw. rückwärts genommenen Differenzen diskretisiert:

$$\operatorname{div}_h \underline{U}_{jk} = \frac{1}{h} (\nabla_x U_{jk} + \nabla_y V_{jk})$$

$$\operatorname{grad}_h W_{jk} = \frac{1}{h} (\bar{\nabla}_x W_{jk} \, , \, \bar{\nabla}_y W_{jk}) \quad .$$

Das finite Problem lautet dann: Gesucht sind Gitterfunktionen U und V sowie $\lambda^{(h)} \in R$ als Lösung von

$$\begin{aligned} \Delta_h \underline{U}_{jk} + c \operatorname{grad}_h \operatorname{div}_h \underline{U}_{jk} + \lambda^{(h)} \underline{U}_{jk} &= 0 \quad \text{auf } \Gamma_h \, , \\ U_{jk} &= 0 \quad \text{außerhalb } \Gamma_h' \, , \\ V_{jk} &= 0 \quad \text{außerhalb } \Gamma_h'' \, . \end{aligned} \tag{6}$$

Der zugehörige Rayleighsche Quotient lautet in Analogie zu (3)

$$R_h(\underline{U}) = \sum_{\Gamma_h} (|\operatorname{grad}_h \underline{U}_{jk}|^2 + c|\operatorname{div}_h \underline{U}_{jk}|^2) / \sum_{\Gamma_h} |\underline{U}_{jk}|^2 \qquad ,$$

wobei formal über das ganze Gitter Γ_h summiert wird.

Untere Schranken werden dann wie folgt erhalten:*)

THEOREM. Die Eigenwerte λ_j von (2) und die Eigenwerte $\lambda_j^{(h)}$ $(j=1,2,\ldots,N)$ von (6) seien der Größe nach geordnet und entsprechend ihrer Vielfachheit aufgeführt. Dann bestehen für $j=1,\ldots,N$ die Ungleichungen

$$\frac{\lambda_j^{(h)}}{1+4(\frac{h}{\pi})^2\lambda_j^{(h)}} \leq \lambda_j \quad . \tag{7}$$

3. Beweis des Theorems.

Ausgehend von zwei auf $\bar{G}$ definierten stetigen Funktionen u und v mit Nullrandwerten auf ∂G und mit quadratisch integrierbaren ersten partiellen Ableitungen u_x, u_y, v_x und v_y werden zugehörige Gitterfunktionen U und V wie folgt konstruiert: Zunächst werden u und v mit der Festsetzung $u=v=0$ außerhalb $\bar{G}$ zu stetigen Funktionen auf die ganze x,y-Ebene fortgesetzt. Dann werden zwei Gitterfunktionen $U(t)$ und $V(t)$ definiert, die noch von dem reellen Parameter $t\in[0,h]$ abhängen:

$$\begin{aligned} U_{jk}(t) &= \frac{1}{h}\int_0^h u(x_j+t, y_k+t+\eta)\,d\eta \quad , \\ V_{jk}(t) &= \frac{1}{h}\int_0^h v(x_j+t+\xi, y_k+t)\,d\xi \quad . \end{aligned} \tag{8}$$

*) In [12] wurde ein analoges Resultat für die Eigenwerte des Stokesschen Operators hergeleitet. Allerdings muß der Faktor im Nenner auch dort $4h^2/\pi^2$ lauten (und nicht h^2/π^2).

Nochmalige Mittelbildung ergibt

$$U_{jk} = \frac{1}{h} \int_0^h U_{jk}(t)\,dt = \frac{1}{h^2} \int_{Q'_{jk}} u(x,y)\,dx\,dy \quad ,$$

$$V_{jk} = \frac{1}{h} \int_0^h V_{jk}(t)\,dt = \frac{1}{h^2} \int_{Q''_{jk}} v(x,y)\,dx\,dy \quad . \tag{9}$$

Im folgenden bezeichnet $\underline{U} = (U,V)$ einen Vektor von Gitterfunktionen U und V, die nach (8) und (9) aus Funktionen u und v konstruiert worden sind. Insbesondere ist also $U_{jk} = 0$ außerhalb Γ'_h sowie $V_{jk} = 0$ außerhalb Γ''_h .

Zunächst sollen einige Beziehungen hergeleitet werden, die zwischen den Integralen

$$D(\underline{u}) = \int_G |\mathrm{grad}\,\underline{u}|^2\,dx\,dy \quad , \qquad E(\underline{u}) = \int_G (\mathrm{div}\,\underline{u})^2\,dx\,dy \quad ,$$

$$H(\underline{u}) = \int_G |\underline{u}|^2\,dx\,dy \tag{10}$$

und den finiten Ausdrücken

$$D_h(\underline{U}) = \sum_{\Gamma_h} [(\nabla_x U_{jk})^2 + (\nabla_y U_{jk})^2 + (\nabla_x V_{jk})^2 + (\nabla_y V_{jk})^2] \quad ,$$

$$E_h(\underline{U}) = \sum_{\Gamma_h} (\nabla_x U_{jk} + \nabla_y V_{jk})^2 \quad ,$$

$$H_h(\underline{U}) = h^2 \sum_{\Gamma_h} (U_{jk}^2 + V_{jk}^2) \quad , \tag{11}$$

$$R_h(\underline{U}) = (D_h(\underline{U}) + c\,E_h(\underline{U}))\,/\,H_h(\underline{U})$$

bestehen .

<u>PROPOSITION.</u> Zwischen den Integralen (10) und den finiten Ausdrücken (11) bestehen die Ungleichungen

$$E_h(\underline{U}) \leq E(\underline{u}) \quad , \tag{12}$$

$$D_h(\underline{U}) \leq D(\underline{u}) \quad , \tag{13}$$

$$H_h(\underline{U}) \geq H(\underline{u}) - 4(\tfrac{h}{\pi})^2 \, D(\underline{u}) \quad . \tag{14}$$

Beweis. Bei der Herleitung von (12) wird die spezielle Art der Mittelbildung (8) und (9) ausgenutzt. Unter Anwendung der Schwarzschen Ungleichung folgt zunächst

$$(\nabla_x U_{jk} + \nabla_y V_{jk})^2 = \frac{1}{h^2} \left(\int_0^h [\nabla_x U_{jk}(t) + \nabla_y V_{jk}(t)]dt\right)^2$$
$$\leq \frac{1}{h} \int_0^h [\nabla_x U_{jk}(t) + \nabla_y V_{jk}(t)]^2 dt \quad .$$

Dabei ist nach (8) und dem Satz von Gauß

$$\nabla_x U_{jk}(t) + \nabla_y V_{jk}(t) = \frac{1}{h} \int_{\partial Q_{jk}(t)} \underline{u} \cdot \underline{n} \, ds = \frac{1}{h} \int_{Q_{jk}(t)} \operatorname{div} \underline{u} \, dx \, dy \, ,$$

wobei $Q_{jk}(t)$ das Quadrat $(x_j+t,\ x_j+t+h) \times (y_k+t,\ y_k+t+h)$ bezeichnet und $\underline{n}$ den nach außen gerichteten Normalenvektor auf dem Rand des Gebietes.

Insgesamt folgt dann unter nochmaliger Anwendung der Schwarzschen Ungleichung und Summation über alle Gitterpunkte

$$E_h(\underline{U}) \leq \frac{1}{h} \int_0^h \left(\int_G (\operatorname{div} \underline{u})^2 dx\, dy\right) dt = \int_G (\operatorname{div} \underline{u})^2 dx\, dy \quad .$$

Das ist die Ungleichung (12).

Aus (9) folgt unter zweimaliger Anwendung der Schwarzschen Ungleichung

$$(\nabla_x U_{jk})^2 = (\frac{1}{h^2} \int_{Q'_{jk}} [u(x+h,y) - u(x,y)]dx\,dy)^2$$

$$= (\frac{1}{h^2} \int_{Q'_{jk}} [\int_0^h u_x(x+t,y)dt]dx\,dy)^2$$

$$\leq \frac{1}{h} \int_{Q'_{jk}} [\int_0^h u_x^2(x+t,y)dt]dx\,dy \ .$$

Summation über das ganze Gitter ergibt

$$\sum_{\Gamma_h} (\nabla_x U_{jk})^2 \leq \frac{1}{h} \int_{\mathbb{R}^2} (\int_0^h u_x^2(x+t,y)dt)dx\,dy$$

$$= \frac{1}{h} \int_0^h (\int_G u_x^2(x,y)dx\,dy)\,dt$$

$$= \int_G u_x^2\,dx\,dy \quad .$$

Analoge Ungleichungen gelten bezüglich u_y, v_x und v_y. Daraus folgt (13).

Zur Herleitung der Ungleichung (14) wird eine Schlußweise aus [8] benutzt. Aufgrund von (9) ist

$$\int_{Q'_{jk}} (u(x,y)-U_{jk})dx\,dy = \int_{Q'_{jk}} u\,dx\,dy - h^2 U_{jk} = 0 \quad .$$

Die Funktion $w = u - U_{jk}$ hat also in dem Gebiet Q'_{jk} den Mittelwert null. Daher besteht die Ungleichung

$$\int_{Q'_{jk}} (w_x^2+w_y^2)\,dx\,dy \geq \mu_2 \int_{Q'_{jk}} w^2\,dx\,dy \quad ,$$

wobei μ_2 den zweiten Eigenwert der freien, schwingenden Membran für das Gebiet Q'_{jk} bezeichnet. Daraus folgt nach Summation über das Gitter Γ_h

$$\int_G u^2\,dx\,dy - h^2 \sum_{\Gamma_h} U_{jk}^2 \leq \frac{1}{\mu_2} \int_G (u_x^2+u_y^2)\,dx\,dy$$

und analog

$$\int_G v^2\,dx\,dy - h^2 \sum_{\Gamma_h} V_{jk}^2 \leq \frac{1}{\mu_2} \int_G (v_x^2+v_y^2)\,dx\,dy \ .$$

μ_2 läßt sich nach unten abschätzen durch $\pi^2/4h^2$. Daraus folgt die Ungleichung (14).

Die untere Schranke für μ_2 läßt sich wie folgt begründen: Ohne Beschränkung der Allgemeinheit betrachten wir etwa das Gebiet $\bar{Q}'_{oo}$, das gegeben ist durch

$$x = t \ , \quad y = t+\eta \ , \quad 0 \leq t \leq h \ , \quad 0 \leq \eta \leq h \quad .$$

Es bezeichne nun w eine Eigenfunktion für das Gebiet Q'_{oo} zum zweiten Eigenwert μ_2. Sie hat den Mittelwert null. Durch Spiegelung an den beiden schrägen Randstücken setzen wir die Funktion w zu einer auf dem Rechteck $\Omega = (0,h) \times (0,2h)$ definierten Funktion $\tilde{w}$ fort. Sie hat auf Ω den Mittelwert null.

Ein einfacher Vergleich der Rayleighschen Quotienten, die zu den Gebieten Q'_{oo} bzw. Ω gehören, liefert dann

$$\mu_2 = \int_\Omega (\tilde{w}_x^2 + \tilde{w}_y^2)\,dx\,dy \Big/ \int_\Omega \tilde{w}^2\,dx\,dy \geq \tilde{\mu}_2 = \pi^2/4h^2 \quad .$$

Damit sind alle Ungleichungen (12) - (14) hergeleitet.

Ähnlich wie in [8],[10] folgt aus den Ungleichungen (12),(13) und (14) das Ergebnis (7):

Es sei zunächst $\underline{u}$ Eigenlösung zum kleinsten Eigenwert λ_1 und $\underline{U}$ der zugehörige, nach (8), (9) konstruierte Gittervektor. Ist $\underline{U}$ nicht identisch null, dann folgt aus der Minimaleigenschaft von $\lambda_1^{(h)}$ zunächst $\lambda_1^{(h)} \leq R_h(\underline{U})$ oder auch

$$\lambda_1^{(h)} H_h(\underline{U}) \leq D_h(\underline{U}) + c\, E_h(\underline{U}) .$$

Abschätzung mittels (12),(13) und (14) ergibt dann

$$\lambda_1^{(h)}(1 - 4(\tfrac{h}{\pi})^2 R(\underline{u})) \leq R(\underline{u}) \quad \text{mit } R(\underline{u}) = \lambda_1 .$$

Daraus folgt (7) für $j = 1$. Ist jedoch $\underline{U}$ identisch null, dann hat $\underline{u} = (u,v)$ die Eigenschaft, daß u und v auf allen Teilgebieten Q'_{jk} bzw. Q''_{jk} den Mittelwert null besitzen. Die Rayleighschen Quotienten von u und v, gebildet für die einzelnen Teilgebiete Q'_{jk} bzw. Q''_{jk}, können daher durch $\pi^2 / 4h^2$ nach unten abgeschätzt werden. Summation über die Teilgebiete liefert dann

$$\frac{1}{4} (\frac{\pi}{h})^2 H(\underline{u}) \leq D(\underline{u}) + c\, E(\underline{u})$$

und damit

$$\lambda_1^{(h)} / (1 + 4(\tfrac{h}{\pi})^2 \lambda_1^{(h)}) \leq \frac{1}{4} (\frac{\pi}{h})^2 \leq R(\underline{u}) = \lambda_1 .$$

In beiden Fällen besteht also die Ungleichung (7) für $j = 1$.

Um die Ungleichung (7) für die höheren Eigenwerte herzuleiten, werden linear unabhängige Eigenlösungen $\underline{u}_1,\ldots,\underline{u}_n$ zu den ersten n Eigenwerten $\lambda_1,\ldots,\lambda_n$ betrachtet. Aus einem bekannten Extremalprinzip zur Charakterisierung der Eigenwerte ([3],[4],[5]) folgt zunächst

$$R(c_1\underline{u}_1 + \ldots + c_n\underline{u}_n) \leq \lambda_n \quad \text{für alle } (c_1,\ldots,c_n) \neq 0 .$$

Sind auch die zugehörigen Gittervektoren $\underline{U}_1,\dots,\underline{U}_n$ linear unabhängig, dann ist bekanntlich

$$\lambda_n^{(h)} \leq \max_{(c_1,\dots,c_n)\neq 0} R_h(c_1\underline{U}_1 + \dots + c_n\underline{U}_n) \quad .$$

Mit geeigneter Wahl von $(c_1,\dots,c_n) \neq 0$ bestehen dann für $\underline{U} = c_1\underline{U}_1 + \dots + c_n\underline{U}_n$ und $\underline{u} = c_1\underline{u}_1 + \dots + c_n\underline{u}_n$ gleichzeitig die Ungleichungen

$$\lambda_n^{(h)} \leq R_h(\underline{U}) \quad , \quad R(\underline{u}) \leq \lambda_n \quad .$$

Genau wie oben folgt dann aus (12),(13) und (14) die Ungleichung (7) für $j = n$.

Sind jedoch $\underline{U}_1,\dots,\underline{U}_n$ linear abhängig, dann ist die Gitterfunktion $\underline{U} = c_1\underline{U}_1 + \dots + c_n\underline{U}_n$ identisch null für geeignete Wahl von $(c_1,\dots,c_n) \neq 0$. Analog wie oben erhält man jetzt

$$\lambda_n^{(h)} / \left(1 + 4\left(\tfrac{h}{\pi}\right)^2 \lambda_n^{(h)}\right) \leq \frac{1}{4}\left(\frac{\pi}{h}\right)^2 \leq R(\underline{u}) \leq \lambda_n \quad .$$

In beiden Fällen besteht also die Ungleichung (7) für beliebiges $n > 1$. Damit ist das Theorem vollständig hergeleitet.

LITERATUR

[1] Landau, L.D., Lifschitz, E.M.: Elastizitätstheorie. Akademie-Verlag Berlin 1965.

[2] Fichera, G.: Numerical and quantitative analysis. Pitman 1978.

[3] Collatz, L.: Eigenwertaufgaben mit technischen Anwendungen. Akademische Verlagsgesellschaft Geest & Portig, 2. Aufl. Leipzig 1963.

[4] Gould, S.H.: Variational methods for eigenvalue problems. University of Toronto Press 1957, Oxford University Press 1966.

[5] Weinstein, A. and Stenger, W.: Methods of intermediate problems for eigenvalues. Academic Press 1973.

[6] Bassotti, L.: Calcolo numerico degli autovalori relativi al primo problema dell'elastostatica piana in un quadrato. Riv.Matem.Univ.Parma,s.2, 9 (1968) 221-245.

[7] Weinberger, H.F.: Upper and lower bounds for eigenvalues by finite difference methods. Comm.Pure Appl.Math. 9 (1956) 613-623.

[8] Weinberger, H.F.: Lower bounds for higher eigenvalues by finite difference methods. Pacific J.Math. 8 (1958) 339-368.

[9] Hersch, J.: Equations différentielles et fonctions de cellules. C.R.Acad.Sci.Paris 240 (1955) 1602-1604.

[10] Hersch, J.: Lower bounds for all eigenvalues by cell functions: A refined form of H.F. Weinberger's method. Arch.Rational Mech.Anal. 12 (1963) 361-366.

[11] Kuttler, J.R.: Upper and lower bounds for eigenvalues by finite differences. Pacific J.Math. 35 (1970) 429-440.

[12] Velte, W.: Eigenwertschranken mit finiten Differenzen beim Stokes'schen Eigenwertproblem. ISNM 43 (1979) 176-188.

W. Velte
Institut für Angewandte Mathematik und Statistik
der Universität, Am Hubland

D-8700 Würzburg

ERRATUM

In der oben zitierten Arbeit [12] W. Velte, ISNM 43 (1979), ist in den Abschätzungen (Seite 184, Formeln (14) und (15), Seite 185 und folgende) der Faktor überall wie folgt zu berichtigen:

$$4\,\frac{h^2}{\pi^2} \qquad (\text{statt: } \frac{h^2}{\pi^2}) \quad .$$

ISNM, Vol. 69
Numerical Treatment of
Eigenvalue Problems, Vol. 3

AN ERROR BOUND FOR EIGENVALUE ANALYSIS BY NODAL CONDENSATION

Heinrich Voss

The nodal condensation is an efficient way of reducing the size of eigenvalue problems to manageable proportions. In this note we use a minimax characterization of the eigenvalues of the exactly condensed (nonlinear) eigenvalue problem to estimate the errors of the eigenvalues of the reduced (linear) problem. The eigenvalue approximations can be improved considerably by the use of the Rayleigh functional of the nonlinear problem.

1. Introduction

The dynamic analysis of structures by finite element or finite difference methods often leads to eigenvalue problems of such a magnitude that its eigensolutions are very expensive to obtain. For large eigenvalue problems Guyan [2] and Irons [4] proposed a method to reduce the size of the matrices which is referred to as eigenvalue economization or nodal condensation or simply reduction.

In this reduction process some of the variables (often known as slaves) are eliminated and one obtains an eigenvalue problem for the remaining uneliminated variables (termed masters) which is nonlinear in the eigenparameter. In the engineering literature (with the exception of [5] and [6]) this problem is linearized yielding the so called reduced problem which generally is much smaller than the original problem.

In this note we consider the nonlinear eigenvalue problem. It turns out that a Rayleigh functional p is defined and that some of the smallest eigenvalues can be characterized by a minimax principle with respect to p.

From this we deduce an a priori error estimate of the eigenvalues of the reduced problem which generalizes the asymptotic error estimate given by Thomas [6]. A posteriori error bounds are given in [2].

Since the Rayleigh functional has similar approximation properties as the Rayleigh quotient in the linear case (first order errors in the eigenvector yield only a second order error in the estimated eigenvalue) the eigenvalue approximations from the reduced problem can be improved considerably if they are replaced by the values of the Rayleigh functional at the corresponding eigenvectors. We demonstrate this improvement by an example where the dimension of the problem is reduced from 705 to 117 and the relative error of the eigenvalue approximation is reduced from about 50% to less than 1% for the ten smallest eigenvalues.

2. The condensation process

We consider the linear eigenvalue problem

$$Kx = \lambda Mx \tag{1}$$

where K and M are real symmetric positive definite $n \times n$ matrices. In the dynamic analysis of structures K and M are the stiffness and mass matrices respectively, x is the vector of nodal displacements and $\lambda = \omega^2$, where ω is the angular frequency.

To reduce the number of unknowns the vector x is partitioned as $x^T = (x_m^T, x_s^T)$, where the subscripts m and s refer to masters and slaves respectively, and (1) is rewritten as

$$\begin{pmatrix} K_{mm} & K_{ms} \\ K_{sm} & K_{ss} \end{pmatrix} \begin{pmatrix} x_m \\ x_s \end{pmatrix} = \lambda \begin{pmatrix} M_{mm} & M_{ms} \\ M_{sm} & M_{ss} \end{pmatrix} \begin{pmatrix} x_m \\ x_s \end{pmatrix}. \qquad (2a,b)$$

Solving equation (2b) for x_s and substituting x_s into equation (2a) one obtains the nonlinear eigenvalue problem

$$T(\lambda)u = 0 \qquad (3)$$

where for convenience we replaced x_m by u and where $T(\lambda)$ is given by

$$T(\lambda) = -K_{mm} + \lambda M_{mm} + (K_{ms}-\lambda M_{ms})(K_{ss}-\lambda M_{ss})^{-1}(K_{sm}-\lambda M_{sm}).$$

To simplify this problem we consider the slave eigenvalue problem

$$K_{ss}\phi = \omega M_{ss}\phi. \qquad (4)$$

Let Ω be the diagonal matrix with diagonal terms ω_i (the eigenvalues of (4)) and let Φ be the s×s matrix which contains as columns the corresponding eigenvectors and is normalized by

$$\Phi^T M_{ss}\Phi = I, \quad \Phi^T K_{ss}\Phi = \Omega. \qquad (5)$$

Then $T(\lambda)$ can be rewritten (cf. Petersmann [5] and for a quadratic approximation Thomas [7])

$$T(\lambda) = -K_o + \lambda M_o + SD(\lambda)S^T \qquad (6)$$

where

$$K_o = K_{mm} - K_{ms}K_{ss}^{-1}K_{sm}$$

$$M_o = M_{mm} - K_{ms}K_{ss}^{-1}M_{sm} - M_{ms}K_{ss}^{-1}K_{sm} + K_{ms}K_{ss}^{-1}M_{ss}K_{ss}^{-1}K_{sm}$$

$$S = M_{ms}\Phi - K_{ms}\Phi\Omega^{-1}$$

$$D(\lambda) = \mathrm{diag}(\lambda^2/(\omega_i-\lambda)).$$

K_o and M_o are called the reduced stiffness and mass matrices respectively. Obviously they are real symmetric and positive definite.

Usually in the literature (with the exception of Petersmann [5] and for the special case $M_{ms} = O$ in Ruge [6]) the linearization of (3) with respect to λ

$$K_o u = \lambda M_o u \tag{7}$$

is solved to get approximations to some of the smallest eigenvalues of (1) and the master portions x_m of the corresponding eigenvectors. x_s then can be obtained from equation (2b). Problem (7) is called the reduced problem.

We now have a closer look at the nonlinear problem (3). Let $\underline{\omega} := \min_i \omega_i$ be the minimal eigenvalue of the slave eigenvalue problem and let $J := (O,\underline{\omega})$. For $u \neq O$ we consider the real function

$$f(\cdot,u) : J \to \mathbb{R}, \; f(\lambda,u) := u^T T(\lambda) u.$$

Then for every $\lambda \in J$ and for every $u \neq O$

$$\frac{\partial}{\partial\lambda} f(\lambda,u) = u^T M_o u + u^T S \operatorname{diag}(\lambda(2\omega_i - \lambda)/((\omega_i - \lambda)^2)) S^T u$$

$$\geq u^T M_o u > O. \tag{8}$$

Hence the real equation $f(\lambda,u) = O$ implicitely defines a functional $p : \mathbb{R}^m \supset D(p) \to J$, $f(p(u),u) = O$, which is called the Rayleigh functional. (Note that in the linear case $p(u)$ is exactly the Rayleigh quotient)

Let

$$\lambda_1 \leq \lambda_2 \leq \cdots \leq \lambda_k < \underline{\omega} \leq \lambda_{k+1} \leq \cdots \leq \lambda_n$$

be the eigenvalues of (1). Then $\lambda_1,\ldots,\lambda_k$ are exactly the eigenvalues of T which are contained in J and the following minimax characterization follows from Theorem 2.1 and Theorem 2.9 in [8]:

THEOREM 1

Let H_j be the set of all j dimensional subspaces of R^m and let $\overline{H}_j := \{ V \in H_j : V\setminus\{0\} \subset D(p)\}$.

Then $\overline{H}_k \neq \emptyset$ and for j=1,...,k

$$\lambda_j = \min_{V\in H_j,\ V\cap D(p)\neq\emptyset} \ \max_{u\in V\cap D(p)} p(u) \tag{9}$$

$$= \min_{V\in \overline{H}_j} \ \max_{u\in V\setminus\{0\}} p(u). \tag{10}$$

3. Error analysis of the reduced problem

Because the eigenvalues of the reduced problem (7) and some of the eigenvalues of the original problem (1) are characterized as minimax values in the same space R^m, we can obtain the following error estimate

THEOREM 2

Let $\lambda_1 \leq \ldots \leq \lambda_k$ be the eigenvalues of problem (1) contained in J and let $\overline{\lambda}_1 \leq \ldots \leq \overline{\lambda}_m$ be the eigenvalues of the reduced problem (7).

Then for j=1,...,k

$$0 \leq (\overline{\lambda}_j - \lambda_j)/\lambda_j \leq \lambda_j/(\underline{\omega} - \lambda_j). \tag{11}$$

Proof

Let $R(u) := u^T K_o u / u^T M_o u$ be the Rayleigh quotient of problem (7). Then

$$p(u) \leq R(u) \quad \text{for every } u\in D(p). \tag{12}$$

For $R(u) \geq \underline{\omega}$ (12) is trivial, for $R(u) < \underline{\omega}$ it follows from $f(R(u),u) = u^T SD(R(u))S^T u > 0$ and (8).

For $j\in\{1,\ldots,k\}$ let $\overline{V} := \text{span}\ \{\overline{u}_1,\ldots,\overline{u}_j\}$ where $\overline{u}_i$, i=1,...,m, denotes an eigenvector of (7) corresponding to $\overline{\lambda}_i$.

If $\overline{\lambda}_j \geq \underline{\omega}$ then $\lambda_j \leq \overline{\lambda}_j$ is trivial; otherwise $\overline{u}_j \in D(p)$ and we obtain from (9) and (12)

$$\lambda_j = \min_{V \in H_j,\ V \cap D(p) \neq \emptyset} \ \max_{u \in V \cap D(p)} p(u)$$

$$\leq \max_{u \in \overline{V} \cap D(p)} p(u) \leq \max_{u \in \overline{V} \setminus \{0\}} R(u) = \overline{\lambda}_j$$

which proves the left inequality of (11).

Denote by $V \in \overline{H}_j$ the j dimensional subspace of R^m such that $\lambda_j = \max_{u \in V_o} p(u)$ where $V_o := V \setminus \{0\}$. Then $p(u) \leq \lambda_j$ for every $u \in V_o$ and it follows from (8)

$$-u^T K_o u + \lambda_j u^T M_o u + u^T S D(\lambda_j) S^T u \geq 0.$$

Hence for every $u \in V_o$

$$\lambda_j \geq u^T K_o u / u^T M_o u - u^T S D(\lambda_j) S^T u / u^T M_o u$$

$$\geq \max_{u \in V_o} R(u) - \max_{u \in V_o} u^T S D(\lambda_j) S^T u / u^T M_o u$$

$$\geq \overline{\lambda}_j - \lambda_j^2 / (\underline{\omega} - \lambda_j) \max_{u \in V_o} u^T S S^T u / u^T M_o u. \qquad (13)$$

From (6) and (5) we obtain

$$SS^T = (M_{ms}\Phi - K_{ms}\Phi\Omega^{-1})(\Phi^T M_{sm} - \Omega^{-1}\Phi^T K_{sm})$$

$$= M_{ms}M_{ss}^{-1}M_{sm} - M_{ms}K_{ss}^{-1}K_{sm} - K_{ms}K_{ss}^{-1}M_{sm} + K_{ms}K_{ss}^{-1}M_{ss}K_{ss}^{-1}K_{sm}$$

$$= M_o - M_1$$

where $M_1 := M_{mm} - M_{ms}M_{ss}^{-1}M_{sm}$ is positive definite.

Thus $\max_{u \in V_o} u^T S S^T u / u^T M_o u \leq 1$, and the right inequality of (11) follows from (13).

Obviously for $\lambda_j << \underline{\omega}$ we get from (11) asymptotically $(\lambda_j - \lambda_j)/\lambda_j < \lambda_j/\underline{\omega}$ which is exactly the error "bound" given in Thomas [7].

4. Computational considerations

The eigenvectors of (3) are the stationary vectors of p. Hence as in the linear case a first order error in the eigenvector approximation u yields a second order error in the eigenvalue estimate p(u), and the eigenvalue approximation $\bar{\lambda}_j$ that we get from the reduced problem should be improved considerably if it is replaced by $p(\bar{u}_j)$.
Generally (e.g. if the slaves are chosen by the automatic criterion of Henshell and Ong [3] for selecting the variables to be eliminated) the computation of $p(\bar{u}_j)$ is very expensive because the slave eigenvalue problem has to be solved.
Quite often however the masters can be chosen such that K_{ss} and M_{ss} are block diagonal matrices and (4) splits into a number of small eigenvalue problems. This situation arises e.g. in the dynamic analysis of a structure which is an assamblage of distinct substructures if the masters and slaves are chosen as the displacements at the grid points on the boundaries and in the interiors of the substructures respectively. In this case the coefficients of the rational function

$$f(\lambda,u) = -u^T K_o u + \lambda u^T M_o u + \sum_{i=1}^{s} |(Su)_i|^2 \cdot \lambda^2/(\omega_i - \lambda)$$

can be compiled substructurewise in very much the same way that is known from finite element methods. Hence the Rayleigh functional can be easily evaluated.
We demonstrate the improvement of the eigenvalue estimates by the Rayleigh functional by the following

Example

$$-\Delta u = \lambda u \text{ in } \Omega,\ u = 0 \text{ on } \partial\Omega, \tag{14}$$

where $\Omega = ((-1,1)\times(0,1))\cup((0,1)\times(-1,0])$ is an L-shaped region. We discretized (14) on a square grid with mesh size h=1/16 using the ordinary 5-point-approximation which yields a linear eigenvalue problem with 705 unknowns.
If we decompose Ω into 12 identical substructures each being a square with side length 1/2 and if we choose the masters and slaves as the boundary and interior grid-points of the substructures respectively then the reduced problem is only of dimension 117.
In the picture on the right we demonstrate the partition of Ω into 12 substructures and the choice of masters and slaves for h=1/6.
The slave eigenvalue problem (4) reduces to 12 identical eigenvalue problems with 49 variables (the solution of which is explicitely known).

× : master variables
• : slave variables

The following table gives the ten smallest eigenvalues λ_j of the discrete problem, the relative errors of the corresponding eigenvalues $\overline{\lambda}_j$ of the reduced problem , and the relative errors of $p(\overline{u}_j)$ where $\overline{u}_j$ is the eigenvector corresponding to $\overline{\lambda}_j$.

j	λ_j	$(\bar{\lambda}_j-\lambda_j)/\lambda_j$	$(p(\bar{u}_j)-\lambda_j)/\lambda_j$
1	9.6735	7.98E-2	3.50E-5
2	15.1468	1.20E-1	1.56E-4
3	19.6759	1.52E-1	3.27E-4
4	29.3418	2.09E-1	2.85E-3
5	31.8269	2.42E-1	5.86E-3
6	41.1262	4.74E-1	-5.59E-3
7	44.4025	5.63E-1	-2.45E-3
8,9	48.8116	5.78E-1	0
10	56.2403	5.22E-1	-8.13E-3

The approximations $p(\bar{u}_j)$ can be improved by the methods for nonlinear eigenvalue problems described in[9](actually we computed the λ_j's by this method).
It is not necessary to take into account all eigenvalues of (4) in the evaluation of the Rayleigh functional. The following table gives the relative errors of $p_j(\bar{u}_i)$ where we considered in $f(\lambda,u)$ only the eigenvalues $\omega_{kl}=h^{-2}(4-2\cos(kh\pi)-2\cos(lh\pi))$, $1\leq k,l\leq j$, of each substructure.

i \ j	1	2	3	4	5	6	7
1	6.5E-3	2.8E-3	5.3E-4	3.0E-4	7.4E-5	5.0E-5	3.5E-5
2	1.4E-2	4.0E-3	1.1E-3	5.3E-4	2.4E-4	1.8E-4	1.6E-4
3	2.0E-2	4.8E-3	1.7E-3	7.5E-4	4.5E-4	3.5E-4	3.3E-4
4	4.1E-2	8.6E-3	5.2E-3	3.4E-3	3.1E-3	2.9E-3	2.9E-3
5	4.4E-2	1.2E-2	8.3E-3	6.4E-3	6.1E-3	5.9E-3	5.9E-3
6	1.3E-2	1.3E-3	-4.2E-3	-4.9E-3	-5.5E-3	-5.6E-3	-5.6E-3
7	1.4E-2	4.8E-3	-1.2E-3	-1.8E-3	-2.4E-3	-2.4E-3	-2.5E-3
8,9	2.7E-2	6.4E-3	1.9E-3	6.1E-4	1.7E-4	3.0E-5	0
10	3.8E-2	-3.2E-3	-5.5E-3	-8.0E-3	-8.3E-3	-8.5E-3	-8.5E-3

References

[1] Geradin, M.: Error bounds for eigenvalue analysis by elimination of variables. J. Sound Vib. 19(1971), 111 - 132

[2] Guyan, R.J.: Reduction of stiffness and mass matrices. AIAA J. 3(1965), 380

[3] Henshell, R.D. and J.H. Ong: Automatic masters for eigenvalue economization. Earthquake Engng. Struct. Dyn. 3(1975), 375 - 383

[4] Irons B.: Structural eigenvalue problems: Elimination of unwanted variables. AIAA J. 3(1965), 961 - 962

[5] Petersmann, N.: Eigenfrequenzanalyse von Systemen mit Substrukturen. To appear in ZAMM

[6] Ruge, P.: Schwingungsberechnung zusammengesetzter Systeme durch modale Synthese. Ing. Arch. 52(1982), 177 - 182

[7] Thomas, D.L.: Errors in natural frequency calculations using eigenvalue economization. Internat. J. Numer. Meth. Engng. 18(1982), 1521 - 1527

[8] Voss, H. and B. Werner: A minimax principle for nonlinear eigenvalue problems with applications to nonoverdamped systems. Math. Meth. Appl. Sci. 4(1982), 415 - 424

[9] Voss, H. and B. Werner: Solving sparse nonlinear eigenvalue problems. Submitted to Numer. Math.

Heinrich Voss
FB Mathematik
Universitätsstrasse 3
D-4300 Essen 1
Fed. Rep. of Germany